THE
MIDDLE EAST
STRATEGIC BALANCE
2003–2004

THE
MIDDLE EAST
STRATEGIC BALANCE
2003–2004

Edited by

Shai Feldman and Yiftah S. Shapir

① Challenge to official Israeli positions

② Excellent summaries & Shai Feldman

sussex
ACADEMIC
PRESS

BRIGHTON • PORTLAND

JAFFEE CENTER FOR
STRATEGIC STUDIES

Chapter Texts and Organization of this Volume
© 2004 Sussex Academic Press / Jaffee Center for Strategic Studies.

The right of Shai Feldman and Yiftah S. Shapir to be identified as Editors of this work has been
asserted in accordance with the Copyright, Designs and Patents Act 1988.

2 4 6 8 10 9 7 5 3 1

First published in 2004 in Great Britain by
SUSSEX ACADEMIC PRESS
P.O. Box 2950
Brighton BN2 5SP

and in the United States of America by
SUSSEX ACADEMIC PRESS
920 NE 58th Ave.   Suite 300
Portland, Oregon 97213-3786

British Library Cataloguing in Publication Data
A CIP catalogue record for this book is available from the British Library.

Library of Congress Cataloging-in-Publication Data
The Middle East strategic balance, 2003–2004 / edited by Shai Feldman.
    p.   cm.
    Includes bibliographical references and index.
    ISBN 1-84519-002-5 (hardcover : alk. paper) — ISBN 1-84519-003-3 (pbk. : alk.
paper)
        1.   Middle East—Strategic aspects.   2.   Middle East—Armed Forces.
3. Middle   East—Military policy.   4.   World politics—21st century.   I. Feldman,
Shai, 1950–

UA832.M523   2004
355′.033056—dc22

2004015675
CIP

Typeset and designed by G&G Editorial, Brighton.
Printed by MPG Books, Bodmin, Cornwall.
This book is printed on acid-free paper.

Contents

Contents

PART II Military Forces

Preface

The Middle East Strategic Balance 2003–2004 marks the twentieth year since the Jaffee Center for Strategic Studies launched its annual flagship publication, which provides extensive analysis of recent developments in the region. The current format allows us to review the year's key strategic developments and assess their implications for the countries involved, for the region, and for the world at large.

This volume continues the Jaffee Center's association with Sussex Academic Press, which began with the publication of *After the War in Iraq: Defining the New Strategic Balance* in September 2003. Sussex expedited publication of the book on the Iraq War, and it was indeed among the first analyses of the war and its ramifications to be published. Parts of the present volume continue the assessment that we at the Jaffee Center began immediately after the military campaign in Iraq officially ended, and it is a pleasure to acknowledge Sussex's help in the publication of the new book.

I would like to express my gratitude to the members of the Jaffee Center staff who made the publication of this volume possible, first and foremost, the researchers who composed and critiqued the material that follows. In addition, my thanks to Yiftah Shapir, head of the JCSS Middle East Military Balance project, and his assistants Tamir Magal and Avi Mor for providing the quantitative data presented. More detailed data regarding the inventories of military forces is available online at the Center's website <www.tau.ac.il/jcss/>. Support for the research and keeping the database current, as well as placing this data online, is made possible through proceeds from the Dr. I. B. Burnett Research Fund for Quantitative Analysis of the Arab–Israeli Conflict.

Thanks also to Moshe Grundman, assistant to the Head of JCSS, who coordinated every aspect of completing this volume, and to our resident editor, Judith Rosen, whose diligence and commitment were the key to preparing this book for publication in a timely fashion.

SHAI FELDMAN, HEAD OF JCSS
TEL AVIV, JUNE 2004

Introduction

There is no question that the Middle East has dominated the global agenda in 2003–2004. The region was rife with events that impacted directly on the contemporary arena and left their imprint for the future, both at home and on the international system. Foreign influences joined local players to help weave the regional fabric; motives ranged between defensive and offensive impulses and the desire to stabilize or dislodge. Yet almost inevitably, any action taken even in the name of correcting injustice or regional inadequacies evoked shockwaves that destabilized the situation while they took time to settle.

Chief among the issues that have occupied the world in this period was the 2003 Iraq War and the still-unfolding events centering on the American presence in Iraq. Diplomatic ramifications, global policies, and regional aftershocks continue to emerge, while Iraq itself is challenged by the difficulties of recovery. As this publication goes to press, the United States has just handed over authority to a temporary Iraqi government as the next stage in post-war reconstruction. However, it is clear to Americans, Iraqis, and international observers that the US presence will likely continue for a while, and that a premature withdrawal could leave Iraq in a state of anarchy, unable to provide for its population and vulnerable to terrorism and radical initiatives. At the same time, it seems that much of the violence currently plaguing Iraq is fueled by the very high levels of an American military presence, and therefore this presence will have to end, or at least be substantially reduced.

Also figuring at the top of the international agenda were the war on terrorism launched by the US in the aftermath of September 11, nuclear proliferation, and the ongoing violent Israeli–Palestinian conflict, approaching the end of its fourth year with no resolution in sight. These subjects, joined with the challenges of a post-war Iraq, are essentially – though not exclusively – Middle East issues that have escalated to seize the focus of world attention. Their impact continues to be felt in the region and carries important ramifications for the international arena that will be felt for many years.

Part I of the present volume contains nine analytical chapters that review and evaluate principal developments in the Middle East and issues that are particularly related to Israel's strategic options. The book opens with Ephraim Kam's strategic survey of the Middle East, which examines the major events in the area against the backdrop of the Iraq War. Kam joins an assessment of the prominent United States role in the Middle East with an analysis of critical developments in the region's leading states. As the introductory overview, the chapter touches on certain issues that are explored elsewhere in the book but are necessarily part of the larger picture depicted. In Chapter 2, Mark Heller probes how the Middle East has captivated international attention. Citing recent global initiatives, Heller questions the prospects for international efforts, particularly in the wake of increasing hostility toward the West among Arab states. Chapter 3, by Yoram Schweitzer, reviews the major developments in international terrorism in 2003–2004. Iraq has become a new locus of terrorist activity, both for al-Qaeda and local groups, and has been joined by al-Qaeda targets elsewhere in the Middle East and Europe.

The focus of the next set of chapters is more regional in nature. In Chapter 4, Emily Landau and Ram Erez survey developments in the Middle East regarding weapons of mass destruction. The threat of nuclear proliferation posed especially by Iran has joined other developments to propel the issue to the center of the international agenda. In Chapter 5, Paul Rivlin surveys the economies of leading Middle East states, dwelling especially on how the upsurge in oil revenues and the region's spiraling population have impacted on these economies. Chapter 6, by Shlomo Brom and Yiftah Shapir, zooms in on Egypt to analyze its extensive military buildup over recent years.

The next two chapters shift to the Israeli–Palestinian conflict. In Chapter 7, Shai Feldman examines the strategic crossroads now before Israel. The momentous change in Israeli national consciousness and the consensus on the need to disengage from the Palestinians lie at the heart of his analysis. Chapter 8, by Anat Kurz, explores the dynamics of the violence that has dominated the conflict since September 2000. Kurz pays particular attention to the disintegration of the centralized Palestinian political and security apparatus, a breakdown that challenges both the Palestinians and Israel. In the final chapter, Meir Elran presents a net assessment of the Middle East in mid-2004, focusing on Israel's strategic position within the region at this time.

Closing the first part of the book are three documents on Israel's disengagement plan, which provide the text of Israel's plan for disengagement from the Palestinians and the letters of understanding exchanged between the Bush administration and the Sharon government. Because these texts are the basis of important developments in the Israeli landscape of 2004 and mentioned in several of the analytical chapters, they are reprinted for the reader who wishes to peruse them in their entirety.

Part II of this volume, compiled by Yiftah Shapir, offers a country-by-country review of the region's military forces, including major changes to the orders-of-battle and key components of their force structures.

Thus, this volume sketches a regional picture and explores leading strategic developments of 2003–2004. The concise review of the region's military forces complements the strategic analysis and shows how military capabilities contribute to the Middle East theater. The goal is to impart to the reader an overall grasp of the region at large and insights into what has characterized and shaped this complex area of the world in the period under discussion.

PART ▶▶▶

I

Middle East Strategic Assessment

CHAPTER ➤➤➤

1

The Middle East in Transition: An Overview

EPHRAIM KAM

Since the beginning of 2003 critical developments in the Middle East have revolved around Iraq, from the war itself and extending to US post-war reconstruction efforts. By mid-2004, the internal and regional ramifications of the Iraqi crisis were not yet fully defined, mainly because the American move in Iraq was still at its height and several years will be needed to evaluate to what extent it has succeeded. However, it is already clear that regardless of ultimate results, the American efforts will have far-reaching effects on the future of Iraq itself and its place in the regional and Arab constellation. The American efforts will impact on the security and perhaps the stability of several major countries in the Middle East. Likewise, they will affect the US role in the region and the US potential capability of meeting other regional and global threats to its security and interests.

Along with the Iraqi crisis and its implications, the violent Israeli–Palestinian conflict, toward the end of its fourth year and with no resolution in sight, continued to engage the US, European governments, and several Arab states, as well as the Israelis and Palestinians themselves. Joining a review of these conflict arenas and against their backdrop, this chapter will examine the stability of the major regimes in the Middle East, following the 2003 milestone of Saddam Hussein's deposal. The chapter will close with a look at some key issues facing Israel in mid-2004.

➤ UNITED STATES INVOLVEMENT

Since the end of the 1980s, US policy in the Middle East has undergone a significant transformation: the US has become more willing to increase its military and political involvement in the Gulf in order to counter serious threats to its strategic interests. This shift evolved from two principal reasons. The first is the prominence of the Persian Gulf since the 1970s as a focus of regional instability and as a source of danger to vital US interests, namely, the flow of oil from the Gulf and the security of several US allies.

The stability of the Gulf was undermined by the rise to power of the Islamic regime in Iran, by the aggressive trends of Saddam Hussein's regime in Iraq, and by the dramatic escalation in the regional arms race that quickly moved to non-conventional weapons. The second reason was the collapse of the Soviet Union, which left the US as the sole superpower in the global theater. This allowed the US to wield its military strength in the Middle East, free of its prior concern that any deterioration in relations resulting from regional tensions might lead to a confrontation between the powers.

The new US willingness to intervene in the region with military force was reflected to a limited extent at the end of the Iran–Iraq War, when American forces blocked the Iranian fleet from interfering with freedom of navigation and the flow of oil in the Gulf. It recurred with far greater magnitude during the 1991 Gulf War, when the US, heading a broad coalition, deployed a massive military force to liberate Kuwait of Iraqi occupation and significantly weaken Iraq's strategic and military power under Saddam.

September 11, 2001 accelerated the US impetus toward military involvement in the Middle East, endowing it with a new and different dimension. The terrorist attack changed the US threat perception: while in the past the perception had focused mainly on the military capabilities of a small group of opposing countries, first and foremost the Soviet Union, it became clear to the US that this concept no longer matched the new reality. In the revised analysis, the critical issues were the threats to US vital interests, to its home front, and to its allies, and the understanding that the terrorist threat is omnipresent, is undeterred by US strength, lacks borders and a clear address, and is liable to materialize at any moment. The new global terrorist threat presented by al-Qaeda and other organizations, a fear of the behavior of radical regimes, especially if they acquired nuclear weapons, and concern regarding the proliferation of weapons of mass destruction (WMD), particularly among terrorist organizations, altered the traditional US concepts regarding national security, threats, superiority, deterrence, intelligence, and early warning. In the eyes of the US these changes obligated a new kind of response, in part also because the potential of strategic surprise had increased, particularly when the US was required to act simultaneously on several fronts, some of which were new. Consequently the US assigned top priority to the proactive struggle against terrorism and the proliferation of WMD, and significantly increased its defense expenditure.

The US regarded the Middle East and its Muslim periphery as a region where many of the new threats were clustered. Those who planned and executed the September 11 attacks were Arabs. The region was home to the bases of al-Qaeda and many other terrorist organizations, and also included several radical countries known as supporters of terrorism. Some of the radical countries in the region were striving to acquire WMD, principally nuclear weapons. Two of the three countries assigned to the "axis of evil" were located in the Middle East. Furthermore, September 11 revised the American ideas on its Middle East role. The US had previously assigned priority to maintaining the stability of the Arab regimes associated with it. Even though it evinced interest in establishing democratic processes and upholding human rights in Arab countries, it avoided making concerted efforts in this direction. When the US did actually employ force in the region in the 1991 Gulf War, it intended to destroy the threat created by the invasion of Kuwait and to weaken Iraq's military and economic strength, in order to prevent it from endangering US allies and interests.

The terrorist attack changed this approach. The danger inherent in the fundamentalist terrorist organizations was brought home to the Bush administration and persuaded it of the need to deal with the deeper currents flowing below the veneer of stability in the Muslim countries. It also concluded that in special cases it was permissible and even necessary to employ military force in order to destroy the terrorist organizations' bases, when the vital interests of the US and the West were at issue. This belief was nourished by the Republican neoconservatives, who eyed the Middle East as fertile ground for their political ideology. Their viewpoint joined the perception of American might – with the fewer constraints on it following the collapse of the Soviet Union – to the fear of fundamentalist terrorism, and the consequent drive to implant democracy in countries regarded as "evil." Their approach comprised several elements: the US must employ its power for the aim of promoting peace and democracy in the world; the use of military force to advance supreme strategic interests must be an important and legitimate instrument of American foreign policy; and the US must consider withdrawal from international agreements that circumscribe its ability to apply force. The military campaign in Afghanistan, and the war in Iraq even more so, were the most important demonstrations of neoconservative philosophy at work.

Consequently, under the umbrella of superpower responsibility and aware of the latent threats in the Middle East to US vital interests and to the West, the American administration took upon itself to redraw important elements of the Middle East picture. It assigned priority to confronting radical Islam and destroying the infrastructure of Islamic terrorist organizations. It strove to influence the policy and behavior of radical regimes that represented a danger to Western interests and, in isolated cases, to topple them or replace them, if necessary by the use of force. It worked to prevent the proliferation of WMD in the region and the acquisition of WMD by radical regimes and organizations. In addition, it hoped to resolve the Palestinian crisis in the spirit of President Bush's vision, which included the establishment of a Palestinian state in 2005.

As part of its evolution from a stabilizing influence to a proactive contributor, the American administration decided it must broaden its approach and devote far greater efforts to the issues of politics, economics, and education in the Arab countries in order to "drain the swamps" that allowed the radical terrorist organizations to breed. This new approach was influenced to a great extent by the Arab Human Development Reports, written by Arab experts and published in 2002 and 2003 under UN sponsorship. The reports portrayed the technological and economic backwardness of Arab countries relative to the Western world as caused by the absence of political and economic freedom and dismally low standards of education. Similarly, the American administration regarded the lack of economic opportunities caused by restrictions on the free market in the Arab world and the absence of political openness and civil institutions as responsible for the despair, frustration, hatred, and incitement to violence in the Middle East. These feelings were understood to nurture the growth of terrorist organizations, the strengthening of radical Islam, and anti-American fervor.

Thus the American administration, even before the war in Iraq, began making efforts to change the face of the Middle East. In December 2002 US Secretary of State Colin Powell launched the US Middle East Partnership Initiative (MEPI), which called for political, economic, and educational reforms. Additional momentum infused the subject of political change in the Middle East at the end of 2003, when the US administration began to formulate the Greater Middle East Initiative as a supplement to the war against terrorism. Its starting point was familiar: only a substantive change to Middle East regimes, to their social values, and to the relations between the West and the Arab/Muslim world can reduce the hostility to the US with a consequent reduction of terrorism. Yet the intentions notwithstanding, it was clear to the US that the establishment of democratic governments in Arab countries would be a gradual, difficult, painful, and dangerous process. Even more important, the administration understood that while Western countries could assist in internal reforms, the democratic change must originate within the Arab societies and could not be imposed by external coercion and management.

Not surprisingly, therefore, the American effort to develop political and economic systems in the Arab world has encountered serious difficulties. The Arab regimes have never displayed enthusiasm for processes of accelerated democracy, because of their fear that reform would play into the hands of the opposition and undermine their own stability. The events of 2003 and 2004 seemed even less conducive to advancing such processes. The occupation of Iraq was regarded by many Arabs, including those who supported the deposing of Saddam Hussein, as part of an imperialist American attempt to achieve hegemony in the Middle East and gain control of Iraqi oil. The Bush administration's policy regarding the Israeli–Palestinian issue was deemed clearly Israel-biased and as awarding Israel broad freedom of action in the Palestinian territories. US pressure on Arab countries to step up their war against radical organizations and to develop more open political and economic systems was likened to interference in their internal affairs and therefore attacked throughout the Arab media. Indeed, the major allies of the US in the Arab world, Egypt and Saudi Arabia, have led the opposition to the Greater Middle East Initiative. Even those Arab countries that understood the need for political openness did not believe that the administration's intentions were untainted and thus did not consider the US a legitimate party for reforming Arab government and society. Rather, they dismissed the American steps as embodying an attempt at external intervention designed to change regimes. Finally, European governments, unenthusiastic about supporting US policy in the Middle East, had their own reservations regarding the American initiatives.

These reactions have aroused concern in the American administration over trends in the Arab world, especially against the background of the US-led war against terrorism and the war in Iraq. Some Arab and Muslim countries assisted US intelligence activities aimed at thwarting terrorist attacks and al-Qaeda in particular, whether because of American pressure on them to do so or because they themselves regard terrorism as threatening their security and stability. However, this cooperation was partial at best, either because these countries feared internal escalation or because some of them, such as Syria and Iran, regarded groups such as Hizbollah and the Palestinian Islamic organizations as liberation movements rather than terrorist organizations. All in all, the administration estimated that there was little support in the Arab world for the battle against terrorism, which made it difficult to combat al-Qaeda and similar organizations. Indeed, while al-Qaeda's organizational and financial structure and its ranks of troops and commanders have been seriously harmed, the organization itself has not been defeated. On the contrary, the popular support it received in numerous Arab and Muslim countries enabled it to set up new, more independent networks, to recruit replacements for captured operatives, and to continue to perpetrate serious attacks in Muslim and Western countries.

To this should be added the antipathy in the Arab world toward the US in the wake of the Iraq War. The war fueled opposition to the US not only by Islamic extremists, but apparently also by secular nationalist elements in the Middle East who regarded the war as part of a struggle between the Western and Muslim worlds and a clash of civilizations. Consequently the administration estimated that support for the US has significantly decreased throughout most of the Muslim and Arab world since 2000, and that this trend both strengthens radical Islam and intensifies the pressures on the regimes supporting the US.

These difficulties raised doubts regarding the American prospects of progressing from its previous role of containing threats to the grander role of designing a new Middle East. As of mid-2004, only partial success was achieved in countering the threats. The efforts to form a new, enlightened region remained in a preliminary stage with their eventual success ultimately dependent on many factors, most of which were out of US control.

➤ THE EVOLVING REALITY IN IRAQ

More than a year after the military campaign in Iraq was officially declared over, the key to regional developments and to the status of the US in the Middle East remained in Iraq. The American administration had defined the strategic aim of the war as the replacement of Saddam Hussein's regime by a stable, democratic, moderate regime linked to the US and the West. Embedded here were clear interests of the administration: success would significantly improve the regional and global status of the US; reduce the harsh anti-American criticism of the war leveled in many countries, including the US itself; positively affect its policy in other Middle East-related issues; and strengthen its allies in the Arab and Muslim worlds.

The American efforts to rehabilitate Iraq also formed a test case for the American drive to develop political and economic systems in the Middle East. The establishment of a democratic regime in Iraq would demonstrate that it was possible to do this in other Arab and Muslim countries. Conversely, failure would likely doom similar efforts. The move in Iraq was therefore regarded by the American administration as a pivotal step in changing the face of the Middle East. Moreover, should it fail, the regional and international status of the US and its deterrence capability would be harmed; Iraq might well become a regional center of instability and a breeding ground for terrorist organizations; and radical Islamic elements were liable to be strengthened, with the US and its allies increasingly hard-put to combat them.

As of mid-2004 the balance of rehabilitation efforts had yielded mixed results. On the positive side, the American administration in Iraq has achieved a series of local successes in constructing the civilian infrastructures from the bottom up: schools and educational institutions, hospitals and clinics, roads and bridges, some of the water supply and irrigation systems, and electricity and oil systems have been built or renovated. A new police force has been set up, government offices were partly operational under Iraqi control, and local councils were established and empowered to act. The Iraqi economy began to revive, although the slow pace of recovery has dissatisfied many Iraqis. The economy suffered from high unemployment partly because of the release of hundred of thousands of young people from the army and the security services, and economic activities have been disrupted because of terrorist attacks, lack of security, and numerous acts of robbery and sabotage.

At the national level, in November 2003, the Coalition Provisional Authority, governing Iraq since the war, concluded an agreement with the Iraqi Governing Council, which drew up a timetable for the phased transfer of power to Iraqi hands. This timetable, which extends until the end of 2005, delineates the formulation of temporary and permanent constitutions, the nature of the presence of US forces in Iraq, the complex process of establishing an assembly and the election of ministers for the local government, and the stages of transferring authority to the Iraqi bodies. As part of this timetable, in February 2004 a temporary constitution was approved for Iraq, albeit skirting some of the residual problems.

The next stage was inaugurated on June 28, 2004, when the US, two days ahead of schedule, handed over authority to a sovereign Iraqi interim government, headed by a prominent Shiite, which replaced the former Iraqi Governing Council. The Coalition Provisional Authority ceased to exist, the occupation of Iraq formally ended, and the Iraqis started to govern their own affairs. The US was to continue to provide technical experts to help Iraq's ministries, but these ministries were responsible to the Iraqi prime minister. American forces were to remain in Iraq under American command, with the precise relationship between them and the Iraqi government to be worked out later. The American plan called for free national elections by January 2005, in which the Iraqi people would choose a Transitional National Assembly. The Assembly in turn was to choose a transitional government with executive powers and draft a new constitution, under which Iraq would elect a permanent government by the end of 2005.

A further positive element was that most of the Iraqi leaders regarded the presence of the American military forces in Iraq as a necessary evil, since their premature departure was liable to lead to the collapse of the entire current regime and to civil war. It was clear to most of them that the Iraqis were incapable of constructing a new regime by their own efforts. Consequently the best chance for the rehabilitation of Iraq as an independent, strong country depended on a continued US presence. For this reason many were willing to cooperate with the American administration, at least as long as they saw a chance for the reconstruction of Iraq that would lead to the departure of the American forces. For its part, the Bush administration was committed to a continued military presence and to reconstructive activities in Iraq, and so far Democratic leaders shared the understanding, since premature withdrawal was liable to cause serious damage to US stature.

However, these achievements were balanced by serious problems. The establishment of local governments and the running of government ministries have not created a sufficient basis for setting up a national regime, which has encountered first and foremost the difficulty of finding a common government denominator between the three major ethnic groups in Iraq – Shiites, Sunnis, and Kurds – particularly when internal struggles marked each of the three. A struggle was thus underway in Iraq between and within the various ethnic groups regarding the form that the country will take and the division of power. This debate has centered on fundamental questions such as: what kind of national structure will Iraq assume? Will it be a federation, and if so, of what kind, or will it be a single, unified body? What will be

Islam's role in the state? Will Iraq become a Western democracy, or perhaps a fundamentalist Islamic regime?

In the current situation, the Shiites, who comprise about 60 percent of the Iraqi population, may be the big winners in the post-war turnaround. Most of the Shiites and their moderate leaders were supporting cooperation with the US in order to establish a democracy in Iraq, since a democratic regime would permit them to become the dominant element in the country. The temporary ruling council included representatives of both secular and religious Shiites, with sharp differences of opinion between them. The senior Shiite religious dignitary, Ayatollah Ali Sistani, was considered to be relatively moderate. For many years he has expressed his opposition to the establishment in Iraq of Iran-style rule by religious leaders, since in his opinion clerics should not take an active part in the government but rather act as religious advisors. On the other hand, Sistani had reservations about the November 15 agreement regarding the election process and the establishment of the Iraqi administration. Furthermore, some of the more extreme Shiite leaders opposed cooperation with the US and advocated the establishment of a religious regime in Iraq. Tension therefore existed among the Shiites between the moderate majority and the extremist minority, which was striving to turn Iraq into a theocracy under Shiite rule.

The Kurds saw the war as an historic opportunity to expand and strengthen the area of the Kurdish autonomy in northern Iraq and to increase their influence over the country in general while exploiting their cooperation with the US. From the very beginning it was clear that they would not achieve independence, because of opposition to the division of Iraq by neighboring states, other groups inside the country, and the American administration. Consequently the Kurds' aim was to establish a strong federal structure in Iraq, and in fact the understanding reached was that Iraq will be reconstructed as a federation. However, in this issue also the Kurds were asked to compromise. They demanded that the planned federation would guarantee the relative independence of the Kurdish region and leave them broad autonomy. In practice, the temporary constitution of Iraq granted them the major autonomy functions they had already achieved, but they have no certainty that a compromise will not be reached at their expense, for example, by establishing a federation on a regional and not an ethnic basis. In any case, it will be necessary to solve the problems associated with the Kurdish region, including ensuring the security and control of the oil resources in it.

The Sunnis, despite being a minority, were the major source of support for Saddam Hussein's regime and have been the primary political casualties of the war. Their removal from power and their fear of activities against them by the Shiites and the Kurds has caused them to react violently. They have shown reservations to cooperation with the US, and extremist elements among them were waging terrorist activities against the American coalition forces, those cooperating with the coalition, and the Shiites, with the aim of achieving the withdrawal of the American forces from Iraq. The Sunnis were still in a preliminary stage of political reshaping following the collapse of the Ba'ath party, in which they played a dominant role. However, some of them have realized that boycotting the political process underway will weaken them even more, and have moved toward rapprochement with the Americans and acceptance of local positions of leadership in the Sunni regions.

The conflict of interests between the various power brokers has emphasized that the process of regime building in Iraq is likely to be protracted and difficult. Although the temporary constitution has been signed, the remaining unsolved problems were liable to disrupt the process of constructing the political system. The various ethnic groups were attempting to exploit the process for their own advantage and in the meantime were setting up their own militias, out of fear that the deteriorating situation might lead to civil war. The transfer of power, addressed in the November 15 agreement, was complex and has encountered a variety of obstacles. Since the fall of Saddam's regime, no outstanding and accepted leadership has emerged in Iraq that could mobilize popular support and create a stable government. Most Iraqis do not understand how democracy works, the parties and the institutions of the civilian society are still in their infancy, and extensive education will be required to operate them. The process was liable to produce a regime led by the Shiites, and there was no certainty that the extremist minority inside them will not be dominant. The Iraqi interim government therefore faced intricate challenges: to acquire legitimacy for itself, to organize the election process, to incorporate the Sunnis in the political system, to organize the Kurdish autonomy in a federal structure, and to define the role of Islam in the state. However, the Americans estimate that the real test of the Iraqi government will begin after the transfer of authority to it. At that stage the regime's degree of legitimacy, its stability, and its capability of overcoming domestic conflicts of interests and of developing democratic life will be fully challenged.

All these issues, however, were overshadowed by the problem of terrorism and lack of security in Iraq. As of the spring of 2004 most of the terrorist attacks in Iraq were apparently perpetrated by Saddam regime loyalists – people associated with the army, the security organizations, and the Ba'ath party – with the aim of creating insecurity in the country, undermining the coalition and ousting the American forces from Iraq, and attacking Iraqi elements, especially Shiites, who were willing to cooperate with the US. The US countermeasures, including the capture of Saddam, hit these people hard but did not defeat them. The difficulty of combating terrorism in Iraq increased because of the activities of terrorist organizations from various Muslim countries, including al-Qaeda, who penetrated Iraq and who were motivated by Arab nationalism, extremist religious ideology, and hostility to the US. Their intention was to establish an extremist Islamic movement hostile to the US that would evict the American forces. Among these, the most outstanding organization was Ansar al-Islam, headed by Abu Musab al-Zarkawi. The terrorist organizations succeeded in building operational, intelligence, and logistical capabilities, and their activities were mainly focused in western Iraq, heavily populated by Sunnis. Violence in Iraq also increased with the activities of many armed militias, which numbered thousands of fighters.

Until the spring of 2004 most of the attacks were directed at Iraqis cooperating with the US, or at Shiites. The extent of the terrorist attacks against the American forces was tolerable, even if they inflicted considerable casualties. Still, however, terrorism threatened the day-to-day security of the Iraqis and deterred some of them from cooperating with the US, and similarly raised difficulties in rebuilding the economy. In addition, it was clear to the Americans that terrorism would most likely increase in the future. Any major terrorist attack causing numerous casualties to US troops could strengthen the American public's opposition to the continued presence of a large force in Iraq, especially if Bush's administration were to be replaced by a Democratic one, less obligated to an overall neoconservative agenda.

The deterioration in the situation in Iraq that began in April 2004 reflected an even graver danger: the possibility of a popular Shiite uprising against the American presence in Iraq. The leadership of this initiative was taken by an armed militia known as the Mahadi Army, headed by a young, charismatic Shiite leader, Muqtada al-Sadr, son of the most prominent Shiite spiritual leader in Iraq who was murdered by Saddam Hussein. The militia was relatively small, since most of the Shiite leaders preferred a quiet process that would transfer to the Shiite majority the leading role in a future Iraqi leadership. However, Sadr – and the Americans suspected, with Iranian support – has placed himself at the head of the opposition to the presence of American forces in Iraq and to any form of cooperation with them. Here he has enjoyed support even from Sunni elements. In mid-2004 the opposition led by Sadr did not spread and was not yet a popular uprising, but it had the potential to foster broad opposition to the American occupation. Uncontained, this threat could endanger American control of parts of Iraq and disrupt the overall reconstruction process.

The US encountered great difficulties without being sufficiently prepared for many of them, while others lay outside its control. The operation's chances of achieving most of its objectives depended on three conditions: American determination to continue; sufficient time, in terms of years, to achieve its goals; and above all recruitment of major Iraqi forces to help attain them. However, the time factor is unknown and might be reduced in the event of multi-casualty terrorist attacks; and the internal contradictions and waves of terrorism make it difficult to obtain cooperation from the Iraqis. Consequently in the middle of 2004, the deterioration in the situation instilled among many in the West, including the US, the feeling that the rebuilding of Iraq has become bogged down, and the insufficient progress planted doubts if the US would achieve its objectives. In any case, the American administration was acutely aware of the dangers of leaving Iraq in an unresolved and disorderly state.

➤ **REGIONAL RAMIFICATIONS OF THE IRAQ WAR**

Even in the relatively nascent post-war era, several regional results of the war are already evident. Yet once more, the war in Iraq demonstrated the weakness of the Arab world, a weakness that has continued for more than a generation. The economic slump has plagued most Arab countries since the mid-1980s, and the forecast for the coming years is not optimistic. The GNP of all Arab countries combined is less than that of Spain alone. The population of the twenty-two Arab countries currently numbers 280 million and is expected to reach 410–460 million by 2020, a dramatic increase that clearly will exacerbate the pressure on available economic resources. The GNP per capita in the Arab world in general has frozen, if not declined, and unemployment is approaching 20 percent. The gap between the rich and poor in Arab countries has continued to widen, but the media revolution has allowed the lower classes to become familiar with the wealth enjoyed by many in the world, which increases frustration and feeds their anger against the local governments.

The weakness of the Arab world is also reflected in the political sphere. Since the waning of Nasserism and the pan-Arab movements at the end of the 1960s, particular national interests dominate pan-Arab goals, and states' conflicts of interest overshadow a common viewpoint. As a result of their desire to wield their political strengths, Arab states have found it increasingly difficult to cooperate with one another. Since 1974, Arab countries have been incapable of unifying and forming a military coalition directed against Israel, and in effect this issue has not been on the agenda for many years. They have also been hard-pressed to agree on a common political stance against Israel, including taking joint steps with respect to the Palestinian crisis. The Arab countries alone were unable to solve the crisis of the Iraqi invasion of Kuwait in 1990, despite its grave ramifications regarding their security and the overall Arab constellation, and were forced to call on the US to solve the problem.

The 2003 Iraq War magnified this weakness. The Arab world found itself humiliated by an invasion and occupation of a key country in the region; the overthrow of a major Arab regime; the rapid collapse of the largest Arab army documented by the media almost live and in real time; and its impotence in the face of this crisis. The Arab world had no real influence on the war – before, during, or after it – and the Arab countries were divided in their attitude to the war and its results. Some of them, including Kuwait, Saudi Arabia, and Jordan, hoped to profit from the change in Iraq, while others, such as Syria, worried about the significance of the American move with regard to themselves. However, the majority were concerned by the possibility that a vacuum might be created in Iraq that was liable to radiate instability on its surroundings, and perhaps also by the possibility that the Shiites would become the major source of power in Iraq.

One of the principal expressions of their political weakness was the last minute postponement of the Arab summit conference in Tunisia in March 2004. Prior to the summit, the first to be held since the war in Iraq, differences of opinion emerged between the Arab countries on a range of topics: the US Greater Middle East Initiative, which fostered opposition from some Arab countries that refused to agree to reforms under American pressure; renewal of the Saudi peace initiative and the overall party line regarding the Palestinian crisis; and the failure of the initiative to strengthen the institutions of the Arab League. Arab leaders eventually met in Tunisia two months later. However, the poorly attended gathering culminated in declarations of solidarity that spoke generally of reforms, but were short on specifics and programs.

The immediate, clear result of the war was the disappearance of Iraq as a major military, political, and to a lesser degree economic player from the regional stage. In effect, the war in 2003 completed the process begun in the 1991 Gulf War. Until then Iraq had been one of the most influential countries in terms of regional developments. Since 2003 Iraq has been occupied, it has no government authorized to and capable of making major independent decisions, and its voice is not heard in the regional arena. Iraq's military strength, both conventional and non-conventional, one of its most outstanding characteristics before the Gulf War, has been totally dismantled. Its economic potential is at this stage mainly controlled by the US.

There is no indication that this situation will substantively change in the near future. The US intends to transfer authority gradually to an interim government in Iraq and rehabilitate its economy. But in any case, this will not be the pre-war Iraq. The American vision is of a moderate Iraq that will engage in internal reconstruction and have a pro-American orientation. If the move fails, Iraq is liable to sink into internecine strife and perhaps a civil war. Most important, Iraq's military power, and especially its non-conventional capabilities, cannot possibly recover for years. The American administration intends to build a small Iraqi army of 35,000 troops, and over time the

US will presumably permit Iraq to rebuild some of its military strength, but this will be done under supervision, on a scale that will not threaten other countries, and in particular will not include elements of WMD.

The collapse of Iraq's military power has removed a major threat to several countries in the region: Iran, Saudi Arabia, Kuwait and the small Gulf States, Jordan, and Israel. Iraq had staged major attacks on Iran and Kuwait, participated with significant force in all the major wars between the Arab countries and Israel, and launched large numbers of ballistic missiles at Iran, Saudi Arabia, and Israel, as well as employing chemical weapons against Iran. Obliteration of the Iraqi threat, at least for the foreseeable future, will enable these countries to divert resources to other defense requirements, as well as to non-defense needs.

Iraq's disappearance as a military player from the Middle East arena and, to a lesser extent, the new moderate orientation of Libya may reduce the Middle East arms race. Since the beginning of the 1990s there has been a definite trend over most years of a reduction of arms imports in the Middle East. According to US State Department data of June 2002 (the latest published until now), in 1989 the Middle East took first place in the world for imports of weapons: its countries imported arms totaling about $22 billion, in 1999 prices, representing approximately 31 percent of world arms sales. The Middle East remained the major arms importer in 1999, but the value of the imports dropped to approximately $13.5 billion, or 26 percent of world arms sales. The reasons for the decrease were the end of the Iraq–Iran War and the cessation of arms imports by Iraq after the Gulf War; the restrictions on the export of arms to Iran; the economic crisis experienced by the Arab countries; the end of arms grants to its clients by the former Soviet Union; and reduction of the threat perception in the Israeli–Arab arena. With the disappearance of the Iraqi threat, vulnerable countries will presumably gradually decrease their arms purchases intended to counter this arrested threat.

The deployment of a large American military force in Iraq alongside smaller forces from other countries is a significant outcome of the war. Still not clear is how long foreign forces will remain in Iraq. The administration intends to keep its forces in Iraq for at least two to three years, but clearly progress toward the administration's goals as well as terrorism and casualty rates will affect their future deployment. In any case, the presence of the American force in Iraq will intensify the pressure on the two neighboring radical countries, Syria and Iran, but will also make it a target for terrorist attacks and a focus for hatred by the Arab world.

The American operation in Iraq brought heavy pressure to bear on the radical regimes in the region – Iran, Syria, and Libya; put them on the defensive; and prompted them to modify their positions. These changes did not result solely from the war, and they should be regarded in a broader perspective. The weakening and disintegration of the Muslim-Arab radical camp began much earlier than the war in Iraq, in fact, in the second half of the 1970s. A series of developments in the region contributed to this process: Egypt's decision to make peace with Israel, which subsequently led the majority of the Arab world to support a diplomatic approach to Israel; Iraq's ongoing difficulties since its war with Iran; the focus of other radical countries (Algeria, Southern Yemen, and Sudan) on their domestic problems; the collapse of the Soviet Union, which traditionally supported the radical camp; the general weakening of the Arab world; and the disappearance of the Arab military coalition against Israel. Since the 1980s a solid Middle East radical camp has ceased to exist, and there has remained

15

only a group of countries and organizations that have generally not acted in concert. On the eve of the war in Iraq this group numbered four countries – Iran, Iraq, Syria, and Libya – whose common features were anti-American policy; an anti-Israeli approach (and, except for Syria, decisive opposition to a peace process with Israel); involvement in terrorism; and efforts to develop various elements of WMD.

The military operation in Afghanistan and the larger international war against terrorism announced after September 11 thrust terrorism to the top of the global agenda and pressured these four countries to reduce their involvement in terrorism and, in particular, to sever their links with al-Qaeda and like organizations. The war in Iraq significantly increased this pressure from several aspects: it broadcast the message of a continued struggle against terrorism, and also emphasized the issue of WMD held by radical regimes; the punishment became tangible, since the US once again applied massive force not only to halt a radical country but also to change its regime, as was done previously in Afghanistan; the change in Iraq was perceived and presented as part of an overall effort to remodel the Middle East with moderate, democratic, and pro-Western regimes in the region; and above all, Iraq, which had been a major member of the radical group, disappeared from the stage as an active player, and thus weakened those remaining in the group.

The most outstanding and significant fallout occurred with Libya. All signs indicate that Qaddafi has taken a strategic decision to change his ways: he revealed his plans for the development of WMD, including those that were not previously known. He has cooperated with the US, Britain, and UN agencies and taken real steps to dismantle the non-conventional weapons program. The change in the Libyan position took place at the end of the 1990s, when Qaddafi sought a means of extricating himself from the isolation and international sanctions imposed upon him because of his involvement in terrorism. The US and Britain offered him a clear-cut deal before the war: if he brought the Pan Am Lockerbie affair to a close and paid substantial compensation to the families of the victims, and if he ceased his involvement in terrorism, the sanctions imposed on Libya by the UN would be removed. If the programs for the development of WMD were also halted, the sanctions imposed by the American administration would be removed and the relations between Libya and the US and Britain would improve. The campaign against terrorism and the war in Iraq accelerated the process by dramatizing the threat to Libya, thus bringing Qaddafi to his decision to cease both his involvement in terrorism and the development of WMD. The change in Libya's stance removed it from the ranks of the radical group, and its improved diplomatic and economic relations with Western powers were intended to strengthen its position as a moderate country.

Syria was also subject to heavy pressure as a result of the war on terrorism and the war in Iraq. The overthrow of the Saddam regime, the appearance of the American forces for the first time on Syria's eastern border, and the weakening of the radical Arab camp following Iraq's and Libya's changed posture left only Syria and Iran in this group. US pressure translated into concrete demands by the administration: cease obstructing US efforts in Iraq and close the eastern border against penetration of Iraq by terrorist fighters; halt the aid to terrorist organizations, mainly Hizbollah and Palestinian organizations maintaining headquarters in Damascus; cease chemical weapons production programs; and withdraw forces from Lebanon.

To this should be added the American decision to impose economic sanctions on

Syria. In October–November 2003 Congress ratified legislation by an overwhelming majority to impose economic sanctions on Syria after the latter failed to meet basic US demands. The law prohibited export of certain types of products from the US to Syria and placed restrictions on American investments in Syria, along with additional financial and commercial restrictions. It also required the president to impose at least two out of six possible sanctions on Syria. Although the law included a non-implementation option if Syria's behavior improved, in May 2004 the president decided to put it into effect. In actuality it is doubtful if the sanctions will have any significant impact, since Syria does not receive aid from the US and there is little trade between them. However, the move was intended to signal that Syria is on the American agenda and that additional steps may be taken if it does not change its policy.

However, Syria has apparently assessed that the American threat is not yet hard and fast, since the US is busy in Iraq and does not seek any additional confrontation with Syria beyond the sanctions. Despite the anger in Washington, the American administration is holding talks with Syria and considers al-Qaeda-related intelligence cooperation to be of value. This assumption has given Syria a certain degree of freedom of maneuver that does not require it to make substantive changes to its policy, but only tactical steps to reduce the pressure on it. These steps have included statements on renewing negotiations with Israel from the point at which they broke off; lowering the profile of the Palestinian organizations with offices in Damascus, but without making a real change to their status; relative restraint of Hizbollah; and partial prevention of infiltration to Iraq from Syrian territory by Islamic activists. At the same time, Syria has aimed to improve its regional status by strengthening the rapprochement with other countries in the region, namely, Iran, Egypt, Jordan, and even Turkey.

Syria's position is of importance regarding Hizbollah. Since 2001 Israel has been concerned that Hizbollah would attempt to escalate the military situation on the Israel–Lebanon border, while exploiting the large rocket installations it has constructed in southern Lebanon, Israel's preoccupation with the Palestinian crisis, and Israel's related desire to avoid opening a second front on the Lebanon border. In practice Hizbollah has from time to time taken military steps in the border region, such as firing on IDF outposts or against aircraft, but as of mid-2004 these have remained relatively circumscribed in location and scope. They may indicate that the organization's considerations are influenced by important constraints: Israel's increased freedom of maneuver against southern Lebanon since the withdrawal of the IDF from the region; the Syrian fear of an Israeli response since Syria has been warned by Israel that it would attack targets in Syria if Hizbollah causes deterioration of the situation; and the desire of the Lebanese government and the residents of southern Lebanon to avoid regional escalation. In early 2004 a deal was completed for an Israel–Hizbollah prisoner exchange, in which Israel released hundreds of Palestinian and Lebanese prisoners, including two senior prisoners, in exchange for a former IDF colonel kidnapped by Hizbollah and the bodies of three IDF soldiers. However, the deal did not greatly influence the organization's behavior.

So far Syria's steps have not indicated a real shift in its position along the Libyan model. Syria's actions toward terrorist organizations were more symbolic than reflective of any substantial change. The positive remarks regarding negotiations with Israel did not reveal any additional Syrian flexibility over previous times when it was under pressure. Syria has taken no confidence building steps toward Israel in order to create a more suitable atmosphere for rapprochement; on the contrary, it continues to encourage terrorist activities against Israel. The consistent Syrian position may result from Assad's conviction that there has been no positive change in Israel's posture or that of the American administration, particularly in the wake of sanctions. In such circumstances the conditions have not yet been created for striking a deal of the Libyan kind between the US and Syria. In Syria's view, there is no reason to dismantle its programs for WMD or even to reduce significantly its support for terror against Israel, as long as it has no better way of combating what it regards as the Israeli threat.

➤ PRESSURES ON IRAN

Iran, the other remaining state in the radical camp, represents a far more complex case. For many years Iran has been subject to significant American pressure in the form of economic sanctions and diplomatic efforts to obstruct the supply of arms and Iran's development of WMD. Since the 1980s Iran has perceived itself as subject to an increasing American threat to its security, economy, and strategic situation. This threat intensified since the September 11 terrorist attack, which placed the war on terrorism at the top of the world agenda, and the campaign in Afghanistan, which both over-threw an Islamic regime supporting terror and deployed American forces on Iran's eastern border.

The war freed Iran of the Iraqi threat, but the instability in Iraq may affect the Iranians, because of the absence of a strong government and because of the Kurdish challenge, liable to spread to Iranian territory. In addition, the US has deployed large military forces in neighboring countries both to Iran's east and west with the aim of overthrowing their Arab/Muslim regimes, in order to dismantle WMD and halt involvement in terrorism. Thus Iran has found itself surrounded on virtually all sides by countries associated with the US, including states where American forces are actually deployed. Against this backdrop, the American administration has presented concrete demands to Iran: cease involvement in terrorism, including links with al-Qaeda; halt interference with American activities in Afghanistan and Iraq by way of increasing Iranian influence there; and above all, cease WMD programs, especially in nuclear weapons. However, the American pressure has been reduced to a certain extent by the US involvement in Iraq, and its difficulties in rehabilitating the country and reconstructing the regime. Furthermore, the vacuum created in Iraq has enabled the Iranians to advance their influence among the Iraqi Shiites.

These major changes to Iran's strategic environment joined the unfolding events surrounding Iran's WMD. Beginning in the summer of 2002 the secret Iranian nuclear program has been exposed by the Western intelligence community, Iranian opposition organizations, the International Atomic Energy Agency (IAEA), and the Iranians themselves. It became clear that the Iranians had for many years engaged in the enrichment of uranium using gas centrifuges and lasers and in attempts to manufacture plutonium, and had constructed a series of secret installations for this purpose with Pakistani, North Korean, Russian, and Chinese aid. In these activities the Iranians violated the commitments they made as signatories to the Nuclear Non-Proliferation Treaty (NPT): they attempted to hide construction of the installations, avoided reporting them as obligated by the convention, attempted to prevent inspections by the IAEA, and lied to the IAEA. Most important, the exposure of the program revealed that Iran was far closer than had been estimated to the production of fissile material, the necessary prelude to the development of nuclear weapons.

The exposure of the Iranian program led to pressure on the IAEA by the US to declare that Iran had violated NPT, which would lead to the imposition of sanctions on Iran by the Security Council. Officials in the administration also hinted that it did not rule out military intervention against Iran if diplomatic activities did not produce results. Thus far the IAEA has refused to declare Iran in violation of the NPT, but has published a series of reports strongly criticizing Iran's nuclear activity. This criticism has also reflected the fact that European countries, headed by France, have revised their estimation of the progress of the Iranian program, and have expressed readiness – apparently to Iran's surprise – to join in the pressure on Iran to halt its nuclear program. The disclosure of the violations led the IAEA to present an ultimatum to Iran in which it was required by the end of October 2003 to: reveal its entire nuclear program, halt the enrichment of uranium and the construction of the enrichment plant, strengthen its cooperation with the IAEA, and agree to more intrusive inspections of its nuclear program.

The heavy pressure, resulting from both the American move in Iraq and international pressure on the nuclear project, has forced Iran to come to a decision regarding its future nuclear plans. Against a background of internal controversy Iran made an intermediate decision: it agreed to make concessions in the nuclear issue that would remove international pressure, but would not eliminate the option of advancing the nuclear program. Thus in October 2003 Iran decided to meet the demands of a group of European countries, backed up by American pressure, to reveal additional significant parts of the program, freeze its activities regarding the enrichment of uranium, increase its cooperation with the IAEA, and sign the Additional Protocol, which obligates it to permit closer supervision of its nuclear activities. On the other hand, Iran has announced that freezing enrichment of uranium is temporary and will be renewed in the future, and it is not obligated to halt construction of the centrifugal plant for uranium enrichment. Furthermore, even after October 2003 the friction between Iran and the IAEA continued, when it was demonstrated repeatedly that Iran attempted to continue to conceal parts of its nuclear activities and enrich uranium, as well as shake off the close IAEA scrutiny.

Although Iran has retreated to some degree in the face of international pressure, its stance does not resemble that of Libya. Qaddafi supplied clear signs that his decision to dismantle his WMD programs was comprehensive and final. Iran's decision in this matter was clearly a tactical one, derived from its fear that a continued attempt to conceal the program and avoid inspections will lead to the immediate imposition of sanctions by the Security Council, perhaps followed by American military action against it. Iran decided not to take chances, to gain time, and to postpone the decision regarding the continuation of its nuclear aspirations to a later date. Furthermore, while the Libyan decision is part of an overall strategic volte face that covers all WMD-related elements, the Iranian decision relates to the nuclear issue only, although its chemical and biological weapons and missiles programs are far more advanced than Libya's.

Thus, in the middle of 2004 Iran faced two possible courses of action regarding the nuclear issue. One was to continue its military nuclear program, with the intention of acquiring nuclear weapons. In order to advance along this path, Iran could continue constructing secret installations for the enrichment of uranium and/or the production of plutonium – and perhaps such installations have already been constructed secretly but not yet discovered – and conceal them, while applying the lessons of the previous failure, in the hope that they will not be discovered this time. Another possibility is to withdraw from the NPT and continue to develop nuclear weapons free of international supervision. It was apparently clear to Iran that continuing its nuclear program through either of these two options would exact a very high price, which could include the immediate imposition of sanctions by the Security Council and perhaps even American military pressure. Most of the signs resulting from Iran's behavior in 2003 and 2004 have indicated that it leans toward continuing to advance its nuclear program, although determining the means and timing at a later stage. If this is indeed Iran's preference, through either of these options or a combination, it seems that the exposure of the program, the additional inspections, and the increased international awareness of its intentions and the steps expected to be taken against it have delayed the timetable for Iran's acquisition of nuclear weapons.

Alternatively, Iran might decide to abandon the program, while striking a deal with international bodies, perhaps including the US, which would award significant advantages to Iran. Up to the summer of 2004 there were no signs that this would be Iran's decision, although there would be sound logic for an Iranian position of this sort. First of all, the Iraqi threat, which was the primary motivation for the development of the program, has disappeared. The American threat of an attack on nuclear and strategic installations has become far more significant, but would be removed with the cessation of WMD development. Second, the overall weakening of the radical camp, with the Libyan model fresh in its mind, may over the course of time persuade elements in Iran that it is possible to reach a reasonable deal with the US, particularly when it is clear that such a deal would bring significant advantages to Iran. Third, Iran cannot ignore the changes in its strategic surroundings when American forces are deployed in Iraq and Afghanistan and help shape their regimes. Iran takes keen interest in developments in these countries, and from this aspect may see advantages in rapprochement with the US.

Iran's strategic decision regarding its course of action, which may be taken in the near future, stands to be influenced by four major considerations. The first is the degree of US capability to drum up massive international pressure on Iran, and Iran's estimation that this pressure will lead to military action against its nuclear installations. Second is the capability of the American administration to offer Iran real advantages, mainly economic and strategic, i.e., regarding the future of Iraq and the security of the Gulf and persuading the Iranians that the American threat to them will be significantly reduced. Third, the developments in Iraq: American success in shaping the future regime in Iraq will intensify the pressure on Iran, whereas American withdrawal without attaining its objectives will play into Iran's hands. Finally and perhaps most important is the effect of the internal struggle in Iran and the chances of a change in the balance between the conservative and moderate factions. From this aspect, the victory of the conservatives in the parliamentary elections of February 2004 is not good news for the possibility of significant rapprochement between the US and Iran, since the conservatives regard such rapprochement as conceding one of the most important attributes of the Islamic revolution, liable to lead to the loss of their power.

➤ THE ISRAELI–PALESTINIAN CONFLICT

The Arab–Israeli crisis, particularly its Palestinian dimension, joined the Iraqi theater as a potential major catalyst for regional instability. The ongoing violence between the Palestinians and Israel neared the end of its fourth year with no resolution in sight. There has been a significant decrease in the number of terrorist attacks, but the violence has continued with little indication that the Palestinians had any inclination to suppress the terrorism. The contacts between Israel and the Palestinians have continued virtually without cessation, including at very high levels, but they have not developed into serious negotiations that could lead to an arrangement.

Perhaps even more ominous was the prevalent impression that both sides and international bodies have despaired of reaching an intermediate arrangement, not to mention a permanent settlement, in the foreseeable future. There was no minimal degree of trust between the sides that could permit an advance toward an intermediate arrangement: any trust created between the sides since the Oslo agreement has been destroyed by the intifada. There was no conceptual framework for an overall agreement that would be acceptable to the respective leaderships, and the attempt by former prime minister Ehud Barak to reach a final agreement in one move failed completely. All the parties involved were ostensibly committed to the roadmap, but they have interpreted it in various ways. In the meantime its practical application was very limited, and the gap between the respective positions remained wide.

Many international bodies – and even Arab leaders and some Palestinians – joined the mainstream Israeli position that Arafat did not want, could not effect, and would not permit an overall agreement. Even the hopes that alternative leaders such as Abu Mazen and Abu Ala would lead the Palestinians to negotiations and an agreement no longer existed. The establishment of the Abu Ala government did not significantly change the situation: he has perhaps maneuvered more effectively than his predecessor inside the Palestinian camp, but he has not achieved freedom of action from Arafat. Consequently in Israel there were few counting on a Palestinian leader to break the political freeze or to halt terrorism.

Palestinians shared the perception of a stalemate. There was no expectation that the Israeli government headed by Ariel Sharon or even an alternative one would reach an agreement with the Palestinians in the near future. Among the Palestinians there was significant support for continued terrorism against Israel, including suicide attacks. Likewise, the Palestinian Authority (PA) has abandoned all real attempts on its part to combat the terrorist infrastructure, and has taken no significant steps to prevent terrorist attacks. The terrorist incentive was also boosted by the involvement of Iran and Hizbollah in attacks launched from the Palestinian territories. The PA was functioning partially at best – some of its organizations have collapsed in the wake of the Israeli intervention in the Palestinian territories – and it is hard-pressed to cope with the deteriorating economic situation and with violations of law and order. The PA has come to terms with the fact that the Fatah is now the leading organization perpetrating terrorist attacks inside the Green Line as well as in the territories, although Palestinian voices have been heard calling for cessation of the armed uprising and the return to a non-violent popular struggle.

From time to time attempts were made among the Palestinians, generally with

Egyptian involvement, to reach agreement regarding partial or temporary cessation of the attacks – mainly terrorist attacks inside Israeli territory, or attacks on civilians – or regarding cooperation between the PA and Hamas. However, up to the middle of 2004, such contacts have not matured into agreements, mainly because of Hamas. Egypt's involvement in these contacts, as well as its readiness to take part in stabilizing the security situation after Israeli withdrawal from the Gaza Strip, brought about a closer dialogue between the Israeli and Egyptian governments. Apart from the Egyptian involvement in the attempts to achieve a ceasefire, however, the Arab countries have done little to deal with the crisis or to aid the Palestinians, either because they saw no real chance of an agreement or because they were too busy with other matters, such as the situation in Iraq, and did not fear the development of internal dissent because of the crisis as in the past.

Prior to the war in Iraq and immediately thereafter, there were expectations that the American administration would invest more intensively in the Palestinian crisis, but this did not happen. The Bush administration continued to deal with the crisis and remained committed to the roadmap and the president's two-state vision, but in practice it assumed a waiting position and its involvement remained limited. Apparently conditions were deemed insufficiently ripe for achieving an agreement, in part because Arafat was blocking possible progress in negotiations and Abu Ala lacked the authority to attain it on his own. In addition, the administration's focus on Iraq and the struggle against terrorism overshadowed other issues, especially since the administration did not identify a real danger of the Palestinian crisis escalating into a regional confrontation. Furthermore, the start of the US presidential election campaign, in which Bush's re-election is not guaranteed, reinforced the hesitation of making an intensive effort to resolve the Palestinian crisis.

However, the number of terrorist attacks, beginning in mid-2002 and especially since 2003, has declined sharply. Between mid-2001 and mid-2002, terrorist attacks peaked, with approximately forty attacks a month perpetrated within Israeli territory west of the Green Line; since the beginning of 2003 this figure has dropped to under ten attacks a month. The decrease in the number of terrorist attacks was also reflected in the ratio of the number of suicide attacks executed to those thwarted. Until the spring of 2002, most of the planned suicide attacks were in fact perpetrated; since then the success rate of attacks has decreased sharply, and from the middle of 2003 the security forces have succeeded in thwarting between eight and nine of every ten planned suicide attacks. By May 2004 the focus of the attacks was transferred to the Gaza Strip, in part because of the IDF activities against the terrorism infrastructures.

Apparently the intense blows to the terrorist infrastructure, including killings and arrests of Palestinian commanders and activists, played an important role. These offensive measures were joined by improvement in the quality of intelligence; the effectiveness of the security fences around the Gaza Strip and in Samaria; and enhanced flexibility, speed of response, and coordination between the security branches. The reduction in the number of attacks also reflected the broad political freedom of action the IDF had in the territories, which was minimally circumscribed to reduce collateral damage and to avoid harming Arafat. Even the assassinations of Hamas leaders Sheikh Ahmed Yassin and Abdul Aziz Rantisi in March-April 2004 were not condemned by the American administration.

Against this background there emerged increasing awareness in Israel of an urgent need to put forward a diplomatic channel that could lead to the end of the crisis, or at least to take the initiative to break the diplomatic deadlock. This recognition emerged from several reasons:

- The understanding among a large part of the Israeli public that Israel has the capability of reducing terrorism and violence, but not of eliminating them solely by military means. Consequently the assessment emerged that if Israel did not take the initiative, it was liable to encounter even more difficult situations, and that it was preferable to try to take the lead when the conditions were relatively good.
- Awareness of the problematic demographic balance between Jews and non-Jews in the territories to the west of the Jordan River, whereby in the foreseeable future there will no longer be a Jewish majority there. Such a situation will not permit maintaining both the Jewish and democratic character of the State of Israel. The sensitivity of this issue increased in light of the wider cracks formed in the texture of the relations between Jews and Israeli Arabs territory during the intifada. There has even emerged a willingness among many people in Israel to exchange areas of Israeli territory with concentrations of Israeli Arabs for areas in Judea and Samaria. However, this readiness has encountered categorical rejection by Israeli Arabs.
- The economic depression, one of whose essential causes was the violence between Israel and the Palestinians.

This situation gave birth to a series of ideas and plans aimed at ending the crisis, cultivated among both the Israeli Left and Right. The Left wished to convince by means of proposals such as the Geneva initiative and the Ayalon-Nusseibeh formula that the best solution for Israel's security lay in a joint agreement with the Palestinians, and that on the Palestinian side there were indeed those who desired and were capable of reaching an agreement. The Geneva initiative, however, was rejected by the current government in Israel, was of dubious promise at best to any future non-left wing government, and was shown in polls as unacceptable to the majority of Israelis. Nor was it clear to what extent it would be accepted by the Palestinians. Among the Right, which has also recognized the need to preserve Israel's Jewish and democratic character, ideas have been suggested for establishing a Palestinian state in Jordan or northern Sinai, or for awarding Jordanian or Egyptian political rights to Palestinian residents in the territories. It is clear, however, that these plans cannot form a basis for a diplomatic move.

This sense of a deadlock has also led many people in Israel, mainly those in the center of the political map among both large and small parties, to support unilateral moves that will ensure Israel's security and its vital interests for at least the coming years, until conditions for a permanent agreement between Israel and the Palestinians ripen. In the last two years, significant support has emerged in Israel for two unilateral measures: the security fence between Israelis and Palestinians to prevent – or at least hinder – Palestinian infiltration to Israeli territory; and disengagement from at least some of the territories. Surveys indicated that these two ideas were accepted in principle by the majority of the Israeli public. However, significant differences of opinion existed regarding their scope and method of implementation.

There was broad agreement in Israel that the construction of a physical barrier was an effective and vital means of preventing Palestinian infiltration, be it to perpetrate terrorist attacks or, though less critically, to settle in Arab towns within the Green Line. The fence built around the Gaza Strip has proven its value as an effective obstacle, and in the regions where it has already been constructed around Judea and Samaria, the fence has made penetration by terrorists difficult and has contributed to the reduction of terrorist attacks. At the same time, there has been sharp controversy regarding the route of the fence in Judea and Samaria. Rightist circles, including Prime Minister Sharon, under pressure by settlement leaders, at first rejected the idea of the fence out of concern that a fence was liable to signal that in the future Israel would be prepared to concede the territories east of the fence. It was also liable to create the impression that the security of the settlers was less important than Israelis residing to the west of the fence. Mounting public pressure, however, and the large number of terrorist attacks in Israel persuaded the government in April 2002 to decide on construction of the fence.

Even subsequently, the implementation of the decision was delayed and did not meet the planned timetable. The precise demarcation was also changed several times, due to the government's oscillating between pressure by the settlers and pressure of international public opinion – including the American administration – that regarded the intrusion of the fence into the territories of Judea and Samaria as a political rather than defensive measure. The Palestinians, who vehemently opposed the fence for creating a fait accompli and for impinging on the daily life and property of thousands of Palestinians, succeeded in persuading numerous governments not to award legitimacy to the fence. Construction of the fence has continued, however, albeit accompanied by many confrontations at the building sites between Palestinians, Israeli and foreign leftists, and IDF soldiers.

The proposed unilateral disengagement was far more complex. At the end of 2003 Prime Minister Sharon began gradually publicizing his intention to carry out a unilateral move of partial disengagement from the Palestinian territories. This was accompanied by debates on the issue within the defense establishment, in the Israeli political scene, and in contacts with the American administration regarding the extent of the disengagement and the conditions under which it would be implemented. In April 2004 the disengagement plan was publicized officially, announcing the intention to dismantle the Israeli settlements and military installations in the Gaza Strip and to remove the IDF forces, except for a section in the south of the Strip that the IDF would maintain to deter arms smuggling from Egypt to the Gaza Strip. The plan also involved the dismantling of four small Israeli settlements and several military installations in northern Samaria. In addition, Sharon made a commitment to the American administration to restrict the expansion of the settlements in Judea and Samaria, to evacuate illegal outposts, and to broaden the freedom of movement for Palestinians who were not involved in terrorism. The plan was scheduled for implementation by the end of 2005.

The contacts with the American administration were intended to obtain its backing and that of additional international bodies for the move, and to receive in return promises of support for the Israeli position regarding other issues involved in an agreement with the Palestinians. In a letter published in April 2004, President Bush expressed his support for the disengagement plan and his commitment to object to any plan other than the roadmap, in which the PA undertook to block and destroy the terrorist infrastructures. Bush emphasized the US commitment to Israel's security and to its definition as a Jewish state, and declared that a just solution to the Palestinian refugee problem lay in the establishment of a Palestinian state to which – rather than to Israel – the refugees could return. Bush also stated that in light of the reality on the ground it was unrealistic to expect future negotiations between Israel and the Palestinians to lead to a complete return to the 1949 borders; rather, the final agreement would be based on mutually agreed changes that would reflect the reality in the country. However, Bush's letter was qualified later by administration spokesmen, who did not actually cancel it, even after the disengagement move was halted, but gave it a meaning less favorable to Israel. For its part the PA welcomed the idea of evacuation and regarded it as an achievement. Nonetheless, it was concerned that the move would lead to the strengthening of Hamas in the Gaza Strip and to a domestic confrontation. It was also worried by the related support given by the American administration to Israel, including Bush's projection that the establishment of a Palestinian state in 2005 was not realistic.

Sharon succeeded in recruiting important international support for the disengagement plan and in receiving an American reward for it. The major obstacle the plan has encountered has instead been within Israel. From the very first it was clear that the settlers would oppose the plan to the best of their ability, and that in this they would be supported by the two right wing parties in the coalition. The major difficulty arose in the Likud, Sharon's own party, many of whose members in the government and in the Knesset were opposed to the plan. In order to bypass this opposition Sharon decided to hold a referendum on the plan among party members, apparently convinced that he would receive approval for his plan. However, in the referendum itself, held in May 2004, the opponents achieved a majority of about 60 percent.

Opposition to the plan rested on two major reasons. The first, which motivated the settlers, the rightist parties, and some members of the Likud, was one of principle: opposition to uprooting settlements and evacuating territory in the Land of Israel, when evacuation of the Gaza Strip was regarded as a precursor to the evacuation of settlements in Judea and Samaria. The second reason was more pragmatic: even among those who were ultimately prepared to evacuate the Gaza Strip, there were some who opposed unilateral evacuation without an agreement, i.e., without receiving anything in return from the Palestinians. To these were added arguments related to the impact of the evacuation on the security situation, prior to and after the evacuation.

Some points of concern in this matter were also raised by senior IDF officers. One was that prior to the evacuation competition was likely to develop between the Palestinian organizations around terrorist attacks, to demonstrate that they had forced Israel out of the evacuated territories. The second was that after the evacuation a wave of attacks was possible, mainly from Judea and Samaria, stemming from the Palestinian feelings that terrorism had led to achievements, the disengagement was evidence of Israeli vulnerability, and intensification of terrorism would result in an Israeli withdrawal from Judea and Samaria as well. The third was that evacuation of the Gaza Strip would lead to the strengthening of Hamas there, which would lead in turn to intensification of the armed struggle. The fourth was that even after disengagement from the Gaza Strip Israel was liable to find itself forced to return to it in order to combat the construction of a reinforced terrorist infrastructure, increased arms smuggling, and ongoing terrorist attacks, perhaps at a higher level than known until now.

The vote by the majority of Likud members against the disengagement plan was a political blow to Sharon. It was caused by Sharon's failure to estimate correctly the strength of the opposition to the plan in his party, conduct a convincing campaign on the necessity of the plan, and combat the comprehensive and effective organizational effort made by the settlers to visit the homes of the voters. Sharon found it difficult to act against the majority position in his party, particularly when this position had earned such an impressive majority. Ultimately, a new plan brokered with specifically murky wording allowed sufficient consensus for it to be passed in the cabinet, especially once the two right wing ministers of the National Union party were dismissed by Sharon.

Consequently, in the middle of 2004 the options for emerging from the Israeli–Palestinian crisis were surrounded by uncertainty and vagueness. Disengagement was officially on the government agenda, but its execution promised to be fraught with challenges. The settlers, encouraged by their victory in the referendum, have already made it clear that they will make every effort to deter the government from carrying out the plan, and to make its implementation difficult. Real dialogue with the PA had not been renewed, even though some people in Israel proposed a reassessment regarding possible negotiations. Other ideas, such as that raised in the National Security Council of reaching a regional Israeli-Palestinian-Egyptian agreement, which would include the exchange of territory and the expansion of the Gaza Strip under Palestinian rule, do not seem promising.

➤ THE STABILITY OF MIDDLE EAST REGIMES

The trend of stability in most Middle East regimes that began in the 1970s continued in 2003–2004. Apart from Saddam Hussein's regime, all other regimes in the region have remained unchanged, and any opposition that exists to most regimes does not threaten their survival. A series of factors accounted for this impressive relative stability: the strengthening of the long-term internal legitimacy of these regimes; the weakness of the opposition to existing regimes; the prevalent stable link between the local regimes and their military establishments, which provided them an important safety net; the difficulty in planning a military coup because of the unwieldy size of the armies; and the effectiveness of the internal security forces in protecting the regimes and handling potential incidents of unrest.

However, several factors are liable at some point to threaten the stability of some of the regimes. One is the socio-economic distress caused by the protracted economic crisis in most of the Middle East, the relatively high unemployment, the inefficiency of the government machinery, the gap between the social strata, and the unfulfilled expectations for improvement in the standard of living and political openness, fed also by the easy access to mass media. The second is the radical Islamic movements, which pose a potential threat to some Arab regimes, even though the strengthening of these movements has been halted in most countries since the middle of the 1990s. The third is the advanced age of several Arab rulers. The process of the rise of a new generation of rulers began in 1999 in three Arab countries – Jordan, Syria, and Morocco – and occurred smoothly. There is no guarantee, however, that successions will be equally peaceful elsewhere. The strengthening of the Kurds in the Iraqi system may also cause the Kurdish problem to spill over to Iraq's three neighbors, Turkey, Iran, or Syria; and the deterioration of the situation in Iraq to one of lack of control is liable to turn it into a hothouse for terrorists, who may in turn act against other countries.

The regime in **Egypt** continues to be characterized by internal stability, and hence perhaps the pressing question of succession. In 2005 Husni Mubarak, who will be seventy-seven, will complete his fourth presidential term, totaling twenty-four years in office, and presumably will be elected to another term if his health permits. Indeed, Mubarak has never appointed a deputy or successor, and uncertainty over the next government has aroused a public debate in Egypt regarding the succession, centering mainly on the possibility of Mubarak's son Gamal as the next ruler. Although at one time Mubarak declared that Egypt is not Syria where rule is handed down from father to son, steps were underway that signaled preparing the future path of Gamal Mubarak to the presidency. He has amassed considerable influence, and he may well become the president of Egypt if agreement is reached between the ruling party and the defense establishment. He was appointed to the leadership of the ruling National Democratic Party and to the head of its diplomatic secretariat, and some of his colleagues have been named to key posts in the party. He has appeared in the media, publicly supporting reforms and increased economic and political openness. His father too has promised to expand political liberalization, for example by canceling some of the military orders that restrict political freedom, but the public has regarded these promises skeptically, seen in part as attempts to comply with American demands. Thus despite the regime's ostensible readiness to permit greater political openness, the implementation of political liberalization even under a new ruler would be a gradual, careful process, and the regime would try to make sure that this does not play into the hands of the extremist Islamic movements.

The danger posed to the stability of the regime by Islamic fundamentalism has not vanished, but has been significantly reduced since the middle of the 1990s by the continued combination of stern measures and readiness for rapprochement displayed by the regime. For example, in September 2003 the leader of al-Jama'a al-Islamiyya, Egypt's most militant Islamic organization, was released after twenty-two years in jail, where he was serving a sentence for assistance in planning the murder of President Sadat. He was released not only because he had served his sentence, but also because he was instrumental in the organization's decision in March 1999 to discontinue the use of violence and to observe an unconditional ceasefire. Furthermore, about a thousand Islamic activists, most of them from al-Jama'a al-Islamiyya, were included in the release of 3,000 prisoners to mark the thirtieth anniversary of the October 1973 war. All those released expressed remorse for their activities and pledged to avoid violence. At the same time, the government denied that the prisoner release reflected a change in policy. It emphasized that it would not accept religious political parties, and indeed, the middle of 2003 saw the arrest of more than one hundred Islamic activists.

The dire economic situation, poverty, and unemployment continued to trouble the regime, instilling in the population feelings of despair, frustration, and disappointment with the government. The regime's major concern was the continued drop in the standard of living. Since the beginning of 2003 there was a depreciation of the Egyptian pound, which caused a rise in prices of imported products and subsequently a shortage of several key products. Although these difficulties embody an ongoing potential for instability, the regime has succeeded up to now in preventing outbreaks of unrest caused by the economic difficulties.

The **Syrian** regime as well has remained essentially stable, enjoying support from important sources of power and facing little serious opposition. After more than four years of Bashar Assad's rule, no one questions his leadership, although he is apparently still heavily influenced by specific advisers and he gives the impression among many people in Israel and other Arab states that he is not as judicious and prudent as his father. Early in his presidency he evinced a desire to permit broader political freedom, but his initiative was blocked by a group of senior members of the Ba'ath party and veterans of his father's regime who argued that too rapid an introduction of liberalization would shake the regime. Assad also wished to introduce economic reforms, but here too he has encountered opposition from senior party officials whose support is critical to him.

Syria's economic situation continued to trouble the country. In 2001–2002 there was a slight improvement in the Syrian economy, but in 2003 the GNP failed to increase, in part because of the disruption of trade with Iraq and the flow of oil from Iraq via Syria as a result of the war. Consequently the Syrian regime continued to face serious economic problems: a low standard of living, increased unemployment, limited activity and investments in the private sector, and a decline in oil reserves. There has also been increased urbanization, evident particularly in Damascus, where the population increased from 2 million in 1980 to 5.5 million in 2004, and urban density coupled with economic woes has made the city a fertile breeding ground for radical movements. The forecasts for the coming years were also not encouraging: a low GNP growth rate, perhaps even negative, is expected, and in the absence of an economic lever (such as a peace treaty with Israel, increased ties to the US, structural reforms, or significant oil discoveries) the crisis is liable to intensify.

Domestic violence erupted in March 2004 in the Kurdish region in northern Syria, beginning as a soccer match brawl and turning into a wave of rioting initiated by the Kurds. The rioting spread to other regions of the country, including the Kurdish quarter of Damascus, and included attacks on government and public buildings, which ended with dozens of people killed. The background to the riots was the tension between the Arabs and the Kurds, who number about a million and a half in Syria, and have protested for some time the Syrian regime's suppression of Kurdish expressions of national identity. The unrest was triggered by the war in Iraq, mainly because the political achievements of the Kurds in Iraq encouraged their brothers in Syria to demand equal rights. In order to prevent further deterioration of the situation, the regime avoided harsh measures against the Kurds and attempted rapprochement, and in fact the riots died down after a few days. This unrest did not represent a real and immediate danger to the stability of the regime, but should it spawn future unrest, especially if supplemental to economic difficulties, it may weaken the regime.

In general the regime in **Saudi Arabia** was stable, though faced with mounting challenges. The most serious threat to the regime was the growing strength of the fundamentalist Islamic bodies, demonstrated by a series of terrorist attacks against Saudi and American targets in Saudi Arabia from the middle of the 1990s. Since the Gulf War the regime has been subject to heightened Islamic pressure, fed by the increased presence of the US in the Gulf and in Saudi Arabia itself, although the US has transferred a large part of its forces from Saudi Arabia to Qatar. The strengthening of the Islamic opposition was also a function of socio-economic factors: the decrease in the standard of living in the kingdom, unequal distribution of wealth, the poverty in the lower strata of the population, increased urbanization, and corruption in the administration.

The regime has regarded extremist Islamic organizations, particularly those connected to al-Qaeda, as a direct threat to its stability, and has therefore taken certain steps against them, including attempts to disrupt their financial pipelines. However, fearing that harsh measures would merely increase support for the organizations, the regime has avoided them. After September 11 the regime declared that it would not tolerate support for terrorism and extremist activities, but it has hesitated to conduct an overall repressive campaign against radical Islamic organizations, and its control over the educational system and incitement in the mosques was inadequate. Initial government changes to the educational system and to textbooks to reduce incitement against foreigners aroused much opposition and were regarded as a surrender to American pressure. At the same time, the minimalist steps have created tension between Saudi Arabia and the US: the Americans strongly criticized the lack of determination in the Saudi handling of the extremist organizations, while hostility in Saudi Arabia against the US because of its activities in the Gulf, the war in Iraq, and its stand on the Palestinian-Israeli crisis continued.

The Saudi economy required urgent reform in order to meet the needs of the rapidly increasing population, which over the last three decades has jumped by 350 percent, from 6.8 million in 1973 to 24.3 million in 2003. This steep increase has created pressure on the labor market and has reduced the per capita income, while foreign workers, who form about a quarter of the population, have created an internal social and political problem. The regime is also required to address the excessive reliance on the oil industry with insufficient development of other branches of the economy, and the need to revive the private sector, to attract investments, and to increase the number of jobs. Crown Prince Abdullah, who has actually ruled the kingdom since 1995, has demonstrated his awareness of the importance of reform, but because of internal constraints, partly due to the close relations between the regime and the Islamic establishment, any reform process would clearly extend over a considerable time period. Prince Abdullah has displayed readiness for a limited opening of the political system, as partial consent to the demand of liberal circles in Saudi Arabia and mainly following pressure by the American administration. For example, in October 2003 the Saudi government announced its intention of holding local elections in 2004, and holding partial elections for the advisory council established after the Gulf War that serves as a kind of appointed parliament sans real authority. However, the impression was that such reforms would occur slowly and in measured fashion, in order to prevent alienation of the Islamic establishment.

The problem of succession has cast a shadow over the stability of the Saudi regime for many years. The current generation of rulers is elderly: King Fahd, who has functioned only partially for the last decade, is eighty-three and his two oldest brothers, Prince Abdullah and Prince Sultan, are eighty and seventy-six, respectively. Their successors will come from a new generation, younger by more than twenty years. Even if the transfer of power is carried out smoothly, the new rulers will have to strike a fresh balance between the traditional religious and cultural values, an outdated and burdened political system, and a conservative economic system versus the challenges of modernity and a changing society. Until now the regime has demonstrated an ability to maintain such a balance. However, in recent years these challenges have been intensified by the extremist Islamic threat, the changes to the strategic surroundings following the war in Iraq, and the pressure by the US on Saudi Arabia to deal in depth with the roots of radicalism in the kingdom. These challenges require the Saudi leadership, both current and future, to introduce reforms, gradually and carefully, into the political and economic systems, in order to adapt the country to the new conditions without shaking its foundations.

The regime in **Jordan** continues to be stable. King Abdullah has enjoyed popularity and support from the major sources of power in the kingdom, mainly the army, the security forces, and the tribal leaders, even if he has not yet established for himself the special personal status in internal and Arab affairs enjoyed by his father. Most opposition elements regard themselves as loyal to the country and have made no efforts to bring down the regime, even if critical of its policy. Jordan has for several years been subject to an economic depression. After the economic crisis that was experienced by most Arab countries since the middle of the 1980s, Jordan enjoyed increased economic growth in the period 1992–1995. However, since 1996–1998 the growth rate of the per capita GNP began to decrease, exports declined, and unemployment increased. Tourism to Jordan declined in the wake of the Israeli–Palestinian crisis and the Iraq War, which also contributed to the recession.

The events occurring to Jordan's east and west have influenced its strategic surroundings. Some of Jordan's population is hostile toward the US and angry over its occupation of Iraq, and the turbulence in Iraq has harmed trade with Jordan. On the other hand, the elimination of the Saddam regime has removed a potential threat from Jordan, and successful reconstruction of the Iraqi economy will also benefit Jordan. In addition, Jordan received the largest grant awarded by the US to its allies in the war – more than a billion dollars of economic and military aid.

Identification with the Palestinians west of the Jordan River is strong among the Palestinian population, and therefore they have expressed anger that the Kingdom of Jordan has done little to help the Palestinians and has not frozen its relations with Israel. This support for the Palestinians has until now not caused internal shocks in Jordan as the regime feared at the beginning of the crisis, but the increasing hostility to Israel within this large sector has forced the king to lower the profile of the relations with Israel, even if diplomacy continues at high levels. For years the regime has been aware of the danger that increasing identification with the Palestinians in the territories presents to its stability. Out of this consideration, in October 2002 King Abdullah launched the initiative "Jordan first," which focused on internal reform and modernization in areas such as the school curriculum, greater freedom of the press, and increased representation of women in the parliament, all with the aim of strengthening Jordanian identity.

Since the 1980s the Islamic organizations and parties have represented the major opposition movement in Jordan. Most of the Islamic organizations have not tried to overthrow the regime and do not threaten its survival. However, the regime fears their growth, in part because of the economic distress, the war in Iraq and the related anti-American feelings, the Palestinian crisis, and the threat that organizations such as al-Qaeda present to moderate Arab regimes. Conscious of the threat that the radical groups present, the regime has occasionally taken steps against them and arrested extremists. In June 2001 the king dissolved the parliament and announced that new elections would be held in November 2001. In parallel a new election law was passed in July 2001, which benefited candidates from the Bedouin regions at the expense of candidates from densely populated regions, where there was greater support for the Islamic groups. Elections were postponed first from November 2001 to August 2002, and again to the spring of 2003, apparently because of the regime's fear that the Islamic groups' parliamentary power would increase as a result of popular anger at the US and Israel. The elections were finally held in June 2003, and the main opposition party, the Islamic Action Front, received only seventeen out of 110 seats. In these circumstances the Islamic opposition, which has been controlled by moderate personages, will apparently not attempt to pass legislation that is contrary to the regime's wishes.

The major factor that has continued to affect regime stability in Iran is the struggle between the elements striving for reform of the political system and the radical establishment. Periodic waves of unrest, demonstrations, and rioting over the last decade, which reflected real discontent and frustration particularly among young people at the political and economic reality, suggest that a large part of the public, if not the majority, is anxious for change in the Islamic regime. The younger generation, which forms the majority of the population in Iran, grew up during the period of the revolution and is not familiar with the conditions that existed under the Shah. Their frustration derives from two major sources: Iran's difficult socio-economic situation since the revolution, reflected by a high unemployment rate even among educated people; and insufficient license for political organization, freedom of speech, and human and civil rights.

The last wave of rioting took place in June 2003 and lasted about a month, and as in the past was led by students. On this occasion the riots were characterized by verbal attacks against not only regime policy, but also its leaders, including personal attacks against the spiritual leader, Ali Khamenei. Furthermore, some of the attacks were directed against the leader of the moderate group, President Mohammed Khatami, because of disappointment in his hesitant leadership and his inability to advance the reform movement. The regime met the protests with a combination of a firm hand and an attempt at rapprochement. Toward the middle of July the riots died down without having escalated beyond control, and the conservative hold over the government was not shaken. The public at large did not join the students, perhaps because they did not believe that the protest would yield results.

The fading of the student protest reflects a far more extensive failure – that of the reform movement to change the basis of the Iranian political system. The reform movement succeeded in exploiting its popular support in order to obtain a majority in several elections held since 1997 and gain control of the presidency, the parliament, and many municipal and local councils. However, the conservative establishment has exploited its control of the state institutions that are not elected – the army, the Revolutionary Guards, the legal and economic systems – and has obstructed the supporters of the reform. The internal struggle exposed the holes in the reform movement: weak leadership, lack of unification, differences of opinion regarding a variety of issues, hesitancy and flinching from confrontations with the conservatives, and a lack of organization that could translate the movement's popular basis into real political power. In particular there is disappointment with Khatami, who balked at confronting the conservatives directly, either because he did not believe that confrontation would achieve his aims or because he feared that such a confrontation would lead to anarchy, bloodshed, and loss of control.

The weakness of the reform movement has led to its decline since 2003. In the municipal elections held in February 2003, many supporters of reform, skeptical of prospects for political change, failed to come and vote, which enabled the conservatives to gain control of many councils. This pattern recurred in parliamentary elections in February 2004. Prior to these elections the conservative establishment disqualified more than a thousand candidates for parliament. Under pressure from the moderate elements, the conservatives restored to the list of candidates a small number of those disqualified. The disqualification of hundreds of moderate candidates and the failure of many voters to vote because they felt that the results were pre-determined permitted the conservatives to regain control in the new parliament. The victory in the elections can enable them to circumscribe the political freedom and to organize politically to gain control of the last outpost held by the supporters of reform, the presidency, in the 2005 elections. However, the fact that the conservatives would hold all the positions of power would make them responsible for governing the country, and blocking reform will leave a potential for unrest and direct confrontation between the parties.

The struggle between the sides will also be affected by the economic situation. In recent years the economic growth rate in Iran has increased; in the period 2000–2002 it exceeded 5 percent per year. The high growth rate, the highest in Iran in the last decade, stemmed from several factors: the rise in oil prices since 1999; changes in the foreign currency policy that expanded exports in fields other than oil; an arrangement for paying the external debt, leading to reduced limitations on imports; and improvement of the agricultural situation because of climatic conditions since 2001. Nevertheless, unemployment is still high and in the period 2001–2002 was estimated at 16–20 percent. There is also a high rate of inflation, which in the 1990s exceeded 22 percent annually. In recent years inflation dropped from 20 percent in 1999 to 14.5 percent in 2000, and to 11.3 percent in 2001. However, it rose again to 17–18 percent in 2002–2003, partly because of the increase in government expenditure. An overall trend toward recovery in the economy is evident, but the widespread poverty and high unemployment continue to pressure the regime.

The current government in **Turkey** was formed in March 2003 by Recep Tayyip Erdogan, the leader of the Justice and Development Party, which in the November 2002 elections received a decisive majority of 363 of the 550 seats in parliament. The party has Islamic roots and Erdogan has an Islamic background. The formation of the government aroused anxiety in certain circles, mainly in the army, that it would adopt an excessive Islamic line. But so far the government has maintained a secular stance and refrained from adopting a prominent Islamic policy, even though its intention to initiate reforms in the educational system is liable to cause internal friction. The major challenge to the government lies with the army, many of whose officers fear that the Erdogan government is opposed to secularism and modernization. However, as yet no real friction between the sides has emerged. On the contrary, several new appointments made in August 2003 in the upper ranks of the army and to the composition of the National Security Council, controlled by the army, may well reduce the risk of friction with the government, since several of the officers replaced were opposed to some of the government's intentions. Until now the major reservation regarding the government's policy has been voiced by President Sezer, who opposed certain decisions related to internal policy, including education, and returned several bills to parliament.

Since the war in Iraq, Turkey and the US have made major efforts to repair the damage to their relations caused by Turkey's refusal to aid the Americans in the war to the extent they had requested. One of the major issues was the American proposal of July 2003 that Turkey send 10,000 troops to Iraq as part of an international force that would help to stabilize the post-war situation. The Turkish government apparently wished to agree to the proposal, both in order to improve the relations with the US and to achieve a degree of influence over events taking place in Iraq. However, Turkey delayed its decision for several months for a variety of reasons: the decision required approval by the parliament, and the government was not sure that it would be granted. Turkey also demanded that the American administration agree to a plan to oust the remaining activists of the Turkish Kurdish party from northern Iraq. The administration agreed to do so, on condition that Turkey would grant them a pardon of sorts, but the Kurds did not agree to the conditions of the partial pardon offered by the government. Turkey demanded, and this was agreed to, that the Turkish force receive a separate region under Turkish command and that it would be given humanitarian assignments, in order to avoid friction with the Iraqi population. Because of these issues that were the subject of internal debate in Turkey, by the middle of 2004 no government decision had been taken regarding a force in Iraq. This affair indicates that as long as the situation in Iraq remains unstable and Turkey is worried about developments in Kurdistan, additional tensions between Turkey and the US would likely remain.

Another question engaging the attention of the Turkish government was negotiations for Turkey's entry to the European Union (EU), which were progressing very slowly. The legal changes that Turkey undertook met, in its opinion, the requirements of the EU and were also welcomed by the EU. However, the EU stated that it wished to be sure that these reforms will in fact be implemented, mainly in issues such as police torture and the use of the Kurdish language in broadcasts and in the educational system. Without such guarantees it has refused negotiations regarding Turkey's EU membership, scheduled to begin in or immediately after December 2004.

Finally, Turkey has for some time been suffering from terrorist attacks, perpetrated mainly by Kurdish and Islamic groups. Two consecutive grave attacks near two synagogues and the British consulate in Istanbul in November 2003 by an Islamic organization apparently connected to al-Qaeda, in which more than sixty people were killed, highlighted the problem of terrorism in Turkey. The secularization of Muslim Turkey and links with the US, Western countries, and Israel underlay the attacks, and the strengthening of al-Qaeda in moderate Muslim and Western countries has marked Turkey as a real target for the future.

➤ ISRAEL IN THE REGION AT LARGE

The rightist coalition formed in Israel after the January 2003 elections seemed to be more sturdy than the national unity government it had replaced, since it had a stable majority, was more homogeneous, and the small parties of the coalition, often the bane of Israel's electoral system, had a clear interest not to upset the balance. Nevertheless, from the beginning of 2004 two issues began to threaten its cohesiveness: Sharon's unilateral disengagement plan, whose approval by the government immediately prompted the departure of the most right wing cabinet members and might obligate the formation of a coalition with the Labor Party; and allegations of Sharon's involvement in bribery and illegal money transfers. By the summer of 2004 it became clear that the prime minister was not going to be charged with these offenses.

The end of 2003 saw initial indications of the end of the severe economic crisis experienced since the end of 2000, although its first signs were felt as early as 1996. Following several years of steady growth, the Israeli economy suffered a severe recession, spurred by two primary factors: the violent conflict with the Palestinians, and parallel trends in the global economy. According to Bank of Israel data, the damage caused by the intifada up to the end of 2003 totaled between 31 and 40 billion shekels, or 6.2–8 percent of the GNP. The sectors of the economy directly affected by the intifada were agriculture, construction, and tourism, and the damage has spread to other fields, mainly related to investments and private consumption. The crisis has also caused significant increases in expenditure because of increased security costs. The recession was reflected in a halting of economic growth and a slowing down of the economy, mainly in industrial production, commerce, services, and the scope of imports, increasing unemployment, and negative financial effects. By mid-2004 limited evidence of an end to the recession appeared, together with the beginning of recovery. These resulted from the government's economic policy and in particular from budgetary cuts and limited reduction of taxes, and also indirectly from the decrease in terrorism since May 2003.

Developments in the Middle East in 2003–2004 have presented Israel with a set of opportunities, though none risk-free. The results of the war in Iraq awarded Israel strategic advantages, at least some of which will be preserved in all circumstances. The Iraqi military threat, both conventional and non-conventional, has disappeared for the coming years. Even when Iraq is permitted to rebuild some of its military strength, this will not be of a magnitude that will threaten Israel. The American move creates considerable pressure on the remaining radical countries, which has already borne fruit: Iran is under pressure to change its behavior, and the exposure of its military nuclear program is expected to lead to at least a postponement of its acquiring nuclear weapons capability. Libya has decided on a strategic volte face and an abandonment of its programs for the development of WMD and its involvement in terrorism. Syria is also subject to increasing American pressure to change its policy. The struggle against terrorism, to which the American administration has assigned top priority, exerts pressure on organizations such as Hizbollah to restrain their activities and affects the fundraising of organizations such as Hamas and Islamic Jihad. Reduction of the number of radical regimes and the establishment of a moderate regime in Iraq will strengthen the moderate bloc in the region. All these developments present important advantages for Israel.

However, alongside these advantages, the results of the war in Iraq also present risks to Israel. If the US move in Iraq fails and it is forced to withdraw its forces without achieving its objectives, some of the advantages derived by Israel will also erode. The status of the US in the Middle East and in the world, of critical importance to Israel, will be weakened and its capability of deterring radical elements will be adversely affected. Iran is liable to become a major force in the Gulf region, and would attempt to fill the vacuum left by Iraq's weakness. Iraq itself might become a focus for regional instability and perhaps even a hothouse for terrorist organizations, and in the future might be ruled again by a radical regime, possibly under Shiite leadership. Such developments would likely give renewed impetus to radical elements in the region and weaken the bloc of moderate Arab countries, including Jordan and Saudi Arabia.

The Israeli–Palestinian crisis did not come closer to an end in 2003–2004, and Israel faces a decision regarding the Palestinian issue. The planned disengagement from the Gaza Strip was approved in June 2004 following much domestic political turmoil. If and when the disengagement plan is carried out, Israel will face a severe internal challenge, the outcome and implications of which are not yet clear. At stake also is Israel's need to act on its national objectives, address the issue of its character as a Jewish democratic state, and make a decision regarding its policy about the territories conquered in 1967 and its control of the Palestinian population. It will have to decide whether to further unilateral moves, whose advantage lies in the fact that they are independent of the agreement of the Palestinian side. However, there are also significant dangers inherent in such moves, because their very unilateral nature does not obligate the Palestinians as would be the case in an agreement. In the meantime Israel must also address the hatred toward it that has arisen in the Arab and Muslim world in the wake of the confrontation with the Palestinians, and has developed into sharp anti-Semitism in the West, mainly in Europe.

The period 2003–2004 reflected a significant decrease in the scale of Palestinian terrorism against Israel, mainly as a result of the steps taken by Israel. It is difficult to estimate to what degree this relative success will continue, but some of the conditions that led to a reduction of terrorism may continue to be advantageous: further construction of the fence, which proved itself as an effective obstacle to infiltration, despite the damage caused to Israel in the international scene; the freedom of action that the American administration – followed by other governments – awarded Israel in its battle against terrorism, even inside PA territory; and Israel's experience in collecting accurate intelligence to destroy the terrorist infrastructure and thwart planned attacks. On the other hand, the increasing involvement of Iran and Hizbollah in Palestinian terrorism and the fact that organizations such as al-Qaeda have marked Israel as a target, even if until now their direct involvement in terrorism against Israel has been limited, may escalate the situation.

The effects of the intifada, the loss of trust between Israel and the Palestinians, the absence of an agreed framework for negotiations, the gap between the adversaries, the reluctance of the American administration to put its full weight behind effecting an agreement and, mainly, Israel's concentration on the unilateral disengagement move: these factors do not portend well for significant negotiations in the near future that would achieve an agreement between Israel and the PA, except perhaps for some security coordination.

The poor chances of negotiations leading to a diplomatic agreement also apply to Syria. Since the withdrawal of the IDF from southern Lebanon in May 2000 and the noticeable reduction in Hizbollah activities against Israel, Syria has lost its trump card against Israel. Israel is under no pressure to reach a quick agreement with Syria while the Syrian president does not display readiness to show flexibility and significantly modify his position, and most of Israel's attention is directed at the Palestinian issue. Furthermore, with Syria subject to increasing American pressure, the American administration has not displayed special interest in Israeli–Syrian negotiations under the current conditions. In this context Hizbollah remains a factor possessing the potential to cause military escalation, albeit relatively limited in scope, between Israel and Syria, although up to now both Syria and the Hizbollah have displayed restraint and have been deterred from moves liable to lead to real deterioration of the situation.

In conclusion, the developments in 2003–2004 have created the foundation for far-reaching ramifications for Israel, since they derive from a move of historical dimensions, namely, the war in Iraq. Much time will have to pass before the full significance of such developments can be assessed. Their significance is liable to become even more complex, when some of the changes expected in the foreseeable future indeed occur. At the same time, Israel must make a strategic decision on the Palestinian issue.

CHAPTER ➤➤➤

2

The Middle East on the Global Agenda

MARK A. HELLER

At the end of the Cold War, for the first time in nearly 200 years, the Middle East ceased to be a focus of great power rivalry. Ever since the Napoleonic invasion of Egypt in 1798, outside powers had competed for presence, influence, and control of the region, if not for its intrinsic value, then at least as a mechanism to improve their geostrategic competitiveness and as a line of imperial communication to other parts of the world. Throughout the nineteenth century, Germany, Italy, Russia, and even Austria-Hungary sought footholds at the edge or in the heart of the region. All this took place against the backdrop of a persistent Anglo-French struggle for preeminence that brought the two countries to the brink of war in 1898, culminated in their dismemberment of the Ottoman Empire at the close of World War I, and lasted until the liquidation of their own empires after World War II. For another half century after that, the Middle East provided a major arena for the Soviet–American competition that came to an end only when the Soviet Union itself collapsed.

The logical consequence of the end of great power rivalry should have been a reduction in intensity of the political-security interaction between the region and the rest of the world. True, the region remained a critical supplier of world energy needs, and its importance in that regard was likely to spiral as the pace of economic growth in such potential giants as China and India accelerated. But other forms of ongoing interaction were not strictly necessary to underwrite these ties, as the declining political leverage on the Middle Eastern policies of such oil importers as Japan seemed to demonstrate. And the same was true, a fortiori, for non-oil economic relations which were, in any case, rather limited. Contrary to the logic of the end of great power competition, however, the interest of the rest of the world not only did not diminish, it seemed to increase, and governments, the UN and other international organizations, NGOs, and the media continued to devote a disproportionate amount of their time, efforts, resources, and attention to the Middle East.

That may have been partly due to the particular fascination of the Western world with the Holy Land as a primary source of its own religious-cultural identity and to the legacy of the West's problematic history with the Jewish people. These factors arguably gave greater prominence to the Israeli–Arab conflict in Western consciousness than what its intensity or destructiveness relative to other protracted international conflicts might warrant. Hence the need to address this conflict remained, if not as a litmus test of Western leaders aspiring to the role of statesman, then at least as something that they could not ignore. But the continuing preoccupation with the Middle East was certainly not because the region as a whole generated the kind of political-strategic or economic power that could be projected outward in a way that might pose either a significant threat or a significant attraction to others. On the contrary, it was the projection of weakness – of systemic failure – that sucked in outsiders hoping to deflect the consequences of that failure far away from themselves.

The symptoms or manifestations of the failure are apparent in almost every realm. In terms of political openness, economic development, and social equality and mobility, the Middle East or, to be more precise, the Arab countries of the Middle East and North Africa, have for years consistently lagged behind every other geographical and/or cultural region in the world, with the possible exception of sub-Saharan Africa. The wave of democratization, marketization, and globalization that successively swept over Eastern Europe, South America, and South and Southeast Asia has virtually passed it by. These general trends were long familiar to specialists with a particular interest in the region, and some of the apparent consequences were already being felt in the 1990s. But they attracted truly urgent global attention only when the enormity of Muslim alienation from the West, evident in the attacks of September 11 and the widespread indifference if not active sympathy they elicited in the Arab Middle East, prompted a more systematic search for "root causes." This imperative was compounded further when the extent of systemic dysfunction in the Arab Middle East was dramatically documented in Freedom House and United Nations Arab Human Development Reports. These demonstrated the prevalence in the Arab world, not only of repressive government and stagnant economies that left few opportunities for political self-expression or economic advancement, but also of a conspicuous knowledge gap, cultural insularity, and lack of citizen empowerment, especially among women.

Had the widespread individual frustration and resentment generated by these conditions been confined to the region itself, they might be a source of regret though not of particular concern to the rest of the world. But the problems of the region have not been contained. Instead, they have been exported abroad in the guise of large numbers of people moving legally or illegally to Europe in search of prosperity or security, as well as large amounts of capital in search of economic opportunities or protection from the threat of instability or arbitrary confiscation. More recently, and especially since regional governments have more or less succeeded in suppressing Islamist opponents without offering any alternative dynamic of their own, the exports have also taken the form of theorists and practitioners of radical ideologies looking either for foreign bases of operations against their own governments or for alternative targets – the Western societies understood to be the sources of both the moral degeneracy and the political durability of their own regimes.

Concerns about the export to the West of the Middle East's systemic failure were implicit in a variety of experiments in social engineering undertaken initially by Europe which, because of its geographic proximity, was the first destination and target of

Middle Eastern surplus manpower, flight capital, and terrorism (the latter primarily in the context of the Algerian civil war). But the linkage was articulated more explicitly after September 11 by the American administration, and perhaps most dramatically in George W. Bush's 2004 State of the Union Address, in which he asserted that "as long as the Middle East remains a place of tyranny and despair and anger, it will continue to produce men and movements that threaten the safety of America and our friends."

The significance of this linkage was an understanding that the lack of freedom, prosperity, and power leading to a widespread sense of individual relative deprivation was mirrored by a lack of national prosperity, power, and prestige leading to a widespread sense of collective relative deprivation. The result of this understanding was a growing disillusionment with the traditional relationship with the Middle East, in which Western governments dealt with Middle Eastern governments only as foreign policy interlocutors, in accordance with Westphalian rules of *realpolitik*, and remained indifferent to the character of local governance so long as it remained dedicated to the sacred cause of "stability." With the conviction that stability had produced stagnation and that stagnation had produced systemic failure whose manifestations could not be contained, the West, and particularly the United States, concluded that the only way to ward off the threat to itself was to engage directly in the region's politics, economies, and societies in order to stimulate reform or transformation.

➤ THE WESTERN AGENDA

In addition to wooing regional allies as part of the East-West contest, Western involvement in the Middle East had always contained an element of the current transformation agenda. Though primarily motivated by that contest, Western foreign assistance projects were also intended to promote the kind of economic development that was believed to be part of, if not a prerequisite to, broader socio-political modernization. But coherent and comprehensive regional transformation projects had to await the end of the Cold War, when they could ostensibly be disentangled from the narrower geopolitical ambitions of the superpowers.

The first of these projects was the Regional Economic Development Working Group (REDWG), established in 1992 as part of the multilateral track of the Madrid Peace Process. Although the "shepherd" for REDWG was the European Union, the United States, as overall sponsor of the process, took an active interest in all the multilateral working groups, including the economic one, and it was particularly supportive of a parallel endeavor, the Middle East/North Africa economic conferences (MENA). Neither of these activities was specifically dedicated to the cause of domestic transformation per se. Instead, they were intended to complement, supplement, and stimulate the bilateral negotiating tracks that were the primary focus of the process, perhaps in the unstated hope that intensification of regional cooperation, particularly in the economic sphere, would facilitate not only peace negotiations but also greater domestic openness. But rather than fulfill those goals, the regional tracks themselves were held hostage to the bilateral tracks— Syria and Lebanon refused from the outset to participate, and the involvement of others waned as the Israeli–Palestinian track deteriorated in the mid-1990s, though that was far from the only reason for the declining enthusiasm.

As if to compensate for the fading away of REDWG/MENA, in 1995 the European Union launched the second post-Cold War regional initiative, the Euro-Mediterranean Partnership (EMP, known as the Barcelona Process). Barcelona was a European initiative prompted by European fears about the consequences of the exported problems from the southern and eastern Mediterranean, especially immigration of underemployed North African youths. But political sensitivities demanded that it be packaged as a "partnership" aimed at regional cooperation in political/security, economic/financial, and social/cultural affairs (or "baskets"). The overall objective of these baskets was liberalization of political, economic, and social systems – greater democracy, judicial independence, and respect for civil and human rights; promotion of the private economic sector (especially small businesses) and free trade; and stimulation of exchanges among civil societies in order to encourage pluralism and tolerance of diversity – in short, the reduction of restrictions on the free flow of goods, services, capital, ideals, and people.

Such changes, however, ultimately threatened the power of governments and privileged economic, social, cultural, and religious establishments in the south. So while bureaucratic-patrimonial "partners" in the Arab world were prepared to entertain cosmetic changes in order to improve their image in the north and were certainly willing to accept the so-called MEDA funds offered by the EU in order to ease the impact of the economic adjustment needed to make possible a Euro-Med Free Trade Area by 2010, they had no real incentive to undermine their own power bases, and the Europeans had no real ability or willingness to push them very hard. As a result, EMP failed to take off. Within a few years, officials had already become preoccupied with analyses of why the Process was stalled and frequently adopted the convenient explanation put forward by Arab "partners," namely, that as in REDWG/MENA, progress here could not be decoupled from progress in the Arab–Israeli peace process. Whatever the real reason, efforts to "reinvigorate" the Process undertaken almost as soon as the Process was launched produced few results, and the main accomplishment of Barcelona has been an enduring series of conferences and meetings. Moreover, the enlargement of the European Union, along with the adoption of its "Wider Europe" initiative, confers EU membership on or at least a "special relationship" with most of the non-Arab Mediterranean partners, essentially reducing Barcelona to a Euro-Arab dialogue with even less certain future prospects.

In the aftermath of September 11 and with Barcelona in seeming disarray, the United States itself embraced the idea of active commitment to the concept of transformation. The first version of this commitment was the so-called Middle East Partnership Initiative (MEPI), announced as a presidential initiative but actually launched by Secretary of State Colin Powell in 2002 and administered by the State Department. MEPI proposed to encourage reform in the region through programs grouped around four "pillars": economics, politics, education, and the empowerment of women. Notwithstanding the seemingly revolutionary packaging, MEPI actually focused concretely on rather modest projects that were redolent of Barcelona, such as seminars on judicial reform, training in entrepreneurial skills, internships for women in business and micro-enterprise, innovative alternatives for relevant education, and a Regional Finance Corporation to help promote a US–Middle East Free Trade Area.

And if financial obligation is any reflection of real intentions, then the American commitment was equally modest: $29 million in fiscal year 2002, $100 million in 2003, and a request for $145 million in 2004.

➤ THE GREATER MIDDLE EAST INITIATIVE

Even before there was any chance to evaluate the impact of MEPI, the US set its sights even higher by elaborating the concept of a Greater Middle East Initiative (GMEI). The contents of a draft GMEI were leaked to the London-based *al-Hayat* newspaper in early 2004, so prematurely that an authoritative version of the Initiative had still not been released by any US government agency almost half a year later. What was leaked amounted to little more than a rehash of the same anodyne assistance programs already incorporated into MEPI, into Barcelona before that, and into bilateral foreign aid programs long before that – proposals to expand civil society and the private sector, train lawyers and judges, and support human and economic development. Nevertheless, the mere appearance of the idea, on the heels of Bush's State of the Union Address and in the psycho-political atmosphere prevailing in early 2004, provoked a torrent of scorn, abuse, and rejection, not only from within the Arab world but also among European leaders, including even close American allies. Europeans attacked it for being geographically diffuse, programmatically unfocused, and contemptuously unilateralist. Many of them also suggested that it betrayed American arrogance and hypocrisy in attempting to impose American values and systems on others while ignoring what they portrayed as the true concerns of those in the region, especially American intervention in Iraq and American support for Israeli policies in the West Bank and Gaza.

Thus, French President Jacques Chirac, who had lauded both Tunisian President Zayn al-Abedine Bin Ali and Egyptian President Husni Mubarak as avatars of democracy and progress, insisted that "any development or modernization in the region presupposes finding a peaceful solution for the Israeli–Palestinian conflict, because the crisis is at the core of difficulties." His foreign minister, Dominique de Villepin, argued that "we need to work in partnership with the countries of the region. We oppose strategies formulated by a worried West trying to impose ready-made solutions from the outside." Even those who, like German Foreign Minister Joschka Fischer, shared the American diagnosis of the "profound modernization crisis in many parts of the Islamic Arab world," nevertheless warned against a self-defeating paternalistic attitude.

Much more flagrant criticism came from within the region. Construing the GMEI as an attempt to bypass governments in favor of "civil society," Arab leaders rejected the Initiative and promised to resist it as an insult and affront to their sovereignty and national pride. Mubarak, for example, responded to the leak by saying: "Whoever imagines that that it is possible to impose solutions or reform from abroad on any society or region is delusional. All peoples by their nature reject whoever tries to impose ideas on them." Jordanian Foreign Minister Marwan Muasher declared, "Our objective is for this document never to see the light." A year earlier, the same Muasher had responded to external initiatives far less ambitious than GMEI by stating, "A wiser approach would be to respect the ability of Arab countries to take matters into their own hands."

Quasi or unofficial comment was even more brutal. *Al-Ahram* columnist Salaheddin Hafez argued that "there is no difference between what was said by the British, French, Belgian, and Dutch colonizers . . . and what the modern empires are saying." Patrick Seale, perhaps writing to rather than on behalf of the Arabs, suggested in *al-Hayat* that that Arabs "should propose a bargain to the United States: 'Resolve the political problems that plague and distract us – Israeli expansion, the plight of the Palestinians, American armed force at the heart of our region, our still incomplete independence – and we will undertake the necessary reforms of our societies, free from the pressures of war and occupation.'"

True, 9/11 and the emergence of a Western transformation agenda did precipitate an unprecedented Arab intellectual debate about the need for change, a debate that was further invigorated when the Arab and Muslim countries were themselves targeted by anti-Western terrorists claiming to speak and act on their behalf. Even among government officials, there were some who argued that the Arabs, if only on tactical grounds, ought to communicate a more responsive attitude to proposals for political and economic reform, even if they emanated from outside sources, including the United States. Their inability to persuade their more recalcitrant colleagues was one of the major factors in the last-minute postponement of the Arab League Summit scheduled for the end of March 2004 in Tunis. Some private individuals were even more outspoken. Mohamed elGhanam, a former colonel in the Egyptian Police and Interior Ministry, wrote in the *International Herald Tribune* (from the safety of Switzerland) to endorse the American diagnosis, to dismiss the "Palestinian card" as anything other than a vehicle for Arab dictators to distract attention from internal problems, and to advocate a vigorous campaign against authoritarian rulers that would favor the stick over the carrot. And Hassan Barari urged "Yes to Reform" in the *Jordan Times*, insisting that the root cause for rejection of the Initiative by regimes was their fear of "empowering the masses," and that the ongoing Arab–Israeli conflict should not constitute an obstacle to Arab reform and democratization any more than it did to Israeli democracy.

On the whole, however, these were marginal opinions. The overwhelming regional response to the Initiative, both official and unofficial, was so hostile that the administration was forced to adopt a defensive and apologetic posture and to downplay and soft-peddle the GMEI, insisting that it wasn't fully formed, that there would be further consultations with allied and regional governments, that the objective truly was a real partnership, and that the Americans too understood that real reform could not be imposed from outside but could ultimately come only from within.

In light of these responses, plans to present the GMEI to the G-8 Summit in June and then to other high-level multilateral forums appeared to have been put on hold or at best modified to incorporate an even more diluted version, and the Initiative began to look stillborn. If that happened, it would probably signal not only the abortion of a specific policy initiative, but also disillusionment with the general aspiration to promote the cause of reform.

There are several possible explanations for the unfolding events and the growing appreciation of how bleak the prospects for transformation might be. Perhaps the most immediate was the rapid deterioration of the situation in Iraq and the realization that a liberal, tolerant Iraqi democracy that would yield a "knock-on" or demonstration effect in the rest of the region – a major post factum if not pre-war rationale for the invasion of Iraq and the ouster of Saddam Hussein – was unlikely to materialize.

A second major obstacle was Israel and America's identification with it. In terms of abstract logic, there was clearly no link between what Israel (with or without American backing) was allegedly doing to Palestinians and what Arab regimes and social establishments were doing to their own political dissidents, cultural iconoclasts, religious apostates, communal minorities, women, economic entrepreneurs, human rights activists, and homosexual "deviants." Israeli policies had no intrinsic connection with Algeria's attitude to Berber cultural autonomy, Morocco's economic performance, press freedoms in Tunisia, the number of book translations in Libya, the ratio of computers to students in Egyptian classrooms, the status of women in Saudi Arabia, parliamentary autonomy or the status of Kurds in Syria, judicial transparency in Lebanon, or the number of internet servers in Yemen. And there was even less evidence of a logical link between the Palestinian issue and the militant struggles of Muslims further afield on what Samuel Huntington called "the bloody borders" of Islam against those seen to challenge the power and dignity rightfully due to Muslims, a category that includes not only Christians (as in Chechnya and the Philippines) but also Hindus (in Kashmir) and Buddhists (in Thailand), and exists not only where Muslims are dominated by others (as in Serbia), but also where they dominate others (as in Iran, Indonesia, northern Nigeria, and Sudan). But abstract logic is different from political logic. Because hostility to Israel and by extension America is so intense, it can easily be manipulated to discredit anyone showing any receptivity to ideas or proposals emanating from America, on the grounds that they are, wittingly or unwittingly, serving the interests of the United States and Israel alike. Consequently, those who would act to promote the cause of reform from within found themselves on the defensive.

Of course, that kind of logic does suggest a kind of "Israel/transformation" paradox: if the unresolved Israeli–Palestinian conflict is a significant obstacle to reform from within, then incumbent regimes and establishments have little real interest in actively promoting a resolution of the conflict lest that unleash forces that could threaten their own hold on power. It also suggests a reason why similar Barcelona Process ideas for reform emanating from Europe, which could not reasonably be suspected of the pro-Israel posture attributed to the United States, elicited more polite but hardly more compliant responses from the Arab partners.

The final and perhaps most daunting obstacle appeared to be the dawning realization that the United States, like Europe before it, cannot actually propose a viable strategy for transformation because it would ultimately be unable to cajole the regimes and status quo forces into cooperating in their own demise – what establishment would voluntarily agree to give up or even share power? It would likewise be unwilling to launch a frontal assault against them, not only because the prospects of failure were high given the shortage of local allies who shared the Western agenda, but also because the consequences of success might be regrettable given the character of the anti-regime forces that do exist. In other words, the West lacks serious indigenous partners for what it proclaimed as a partnership for transformation. Yet Western powers alone can do little to mobilize civil societies that are weak, fragmented, and lacking in democratic roots.

It is not the case in 2004 that civil society is nonexistent. But many of the NGOs that have been identified as potential allies in the quest for reform were really

GONGOs (Government Owned and Operated NGOs), like the Saudi National Committee on Human Rights, established with much fanfare by the Saudi government in March 2004, about a week before the government arrested thirteen liberal academics and reformers for, among other infractions, demanding the establishment of a truly independent Human Rights Committee. Even those unofficial networks and associations that are independent are not necessarily intrinsically committed to the cause of liberalization and democratization; some are thoroughly apolitical and others are actually dedicated to something quite different from what the West seeks in its transformation plans. Perhaps the most prominent real NGO in the Arab/Muslim world is al-Qaeda.

Nor is it the case that sources of a democratic tradition are totally absent. Egypt, for example, carried out an interesting experiment in constitutional liberalism and parliamentary democracy in the 1930s. Lebanon displayed many characteristics of a liberal social order for years before the civil war, along with indicators of modernity and human development – some of which even persisted during the war or were revived since then. There are other noteworthy pockets of political and social liberalization in some of the modernizing monarchies, such as Jordan, Bahrain, Oman, and Morocco. But in the larger regional scheme of things, these have been transitory phenomena or plants that failed to germinate, and the ultimate challenge, as formulated over a decade ago by a prominent Arab intellectual living in the West, remains that of building democracy without democrats.

If this challenge is perceived as insurmountable, the outcome may eventually be complete disillusionment with transformation as a Western response to Middle Eastern systemic dysfunction and a resort to containment as the preferred defense against the corresponding security threat. In other words, Western powers may conclude that the least unpromising course of action is to abandon engagement and to undertake a kind of unilateral disengagement, that is, to minimize human interaction and isolate and contain parts of the Arab/Muslim world until they, too, are themselves ready to jumpstart transformation, as did Turkey under Kemal Ataturk after World War I.

Of course, there are strict limits on the extent to which the West can cut off or cut itself off from the Middle East, at least as long as it remains dependent on Middle Eastern sources of energy. And whatever security measures are introduced, constraints on the movement of people will also be continually tested by existing family and business ties. Moreover, a policy of disengagement and containment would not address the problem of dealing with the consequences of dysfunction that have already been exported. So while such a policy might well be seen as the best way at least to keep the problem from getting worse, implementing it may be no less daunting or more feasible than is a policy of transformation through engagement. If that turns out to be the case, then the only alternative, and the line of least resistance, will be a reversion to the ad hoc "muddling through" that prevailed before attempts at transformation were launched and that produced such problematic results.

CHAPTER ▶▶▶

3

International Terrorism in the Shadow of the Iraq War

YORAM SCHWEITZER

Terrorism has figured prominently on the international agenda for the last several years, but from 2003 onward it appeared largely against the background of the Iraq War. Suicide terrorism, which increased noticeably in this period in areas outside of Israel and the occupied territories, was now staged in Iraq along with other sites in the Middle East and beyond. Al-Qaeda and its supporters have used the war in Iraq, the ensuing low level conflict, and the presence of foreigners for propaganda purposes, in order to justify the murderous terrorist attacks that they were in fact already committing in places throughout the world. Under cover of the war, while claiming to be "defenders of Islam against the Crusader-Jewish attack" they assert is underway, al-Qaeda and its allies have continued their terrorist attacks in various locations around the world. Countries and assistance organizations that sent forces to Iraq to help maintain law and order together with local law enforcement agencies or to help Iraq's residents in humanitarian projects became a target for terrorist attacks or were threatened with future terrorist attacks if they did not remove their personnel from Iraq. Some attacks have been aimed against Arab regimes deemed infidels and collaborators, especially Saudi Arabia, and some have been a realization of previous plans to attack in Europe. Two waves of terrorist attacks in Turkey and particularly those in Spain horrified Europe and shattered the illusion of security it had previously enjoyed.

The countermeasures taken by the international coalition against terrorism have been bilateral and multilateral efforts mainly in the areas of intelligence, law enforcement, and diplomacy. International cooperation between the security systems taking part in the formidable struggle against al-Qaeda and its allies, especially after the September 11 attacks, has led to impressive tactical achievements, including the apprehension of high level operatives. Nevertheless, it is evident that the ability of al-Qaeda and its allies to carry out deadly terrorist attacks around the world has not been blunted; their capabilities in this respect have been demonstrated anew on various occasions. In this state of affairs, security agencies around the world have grown concerned about mega-terrorism, including attacks using non-conventional materials that al-Qaeda and its affiliates have been trying to manufacture or buy. In May 2003, a Saudi Arabian religious leader published a *fatwa* permitting the use of non-conventional weapons, even if this measure caused the death of hundreds of thousands of infidels. The threat of terrorism, especially non-conventional terrorism, continues to hang like a dark cloud over Western countries, as reflected in occasional statements by the heads of the secret services in the US, UK, and other countries, illustrating that this danger will probably persist for years.[1]

➤ THE SPREAD OF SUICIDE TERRORISM

Although suicide terrorism is not new to the world, it appears to have greatly expanded since early 2003 and has spread to regions where it was previously unknown. The primary increase was in the large number of suicide terrorists operating in Iraq, which until the war had not experienced this brand of terrorism. From the beginning of the war until mid-June 2004, sixty-one suicide attacks involving seventy-nine suicide terrorists and causing over 840 fatalities have taken place in Iraq. In Chechnya, fifteen terrorists, including eleven women, committed eleven suicide attacks since the beginning of 2003, killing about 200 people.[2] Suicide attacks continued to constitute a key terrorism weapon in the Israeli–Palestinian conflict. Despite the high number of suicide attacks thwarted by the Israeli security forces, twenty-six suicide attacks were carried out against Israelis between the beginning of 2003 and mid-June 2004. In other areas around the world, fifty-four terrorists identified with al-Qaeda or its affiliates took part in nineteen suicide attacks in which over 280 people were killed. Overall, more than 170 suicide terrorists carried out 117 suicide attacks, a figure that does not include the numerous frustrated suicide attacks, mostly but not exclusively in Israel (figure 3.1).

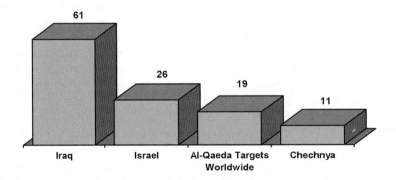

Figure 3.1   Distribution of Suicide Attacks by Country, Jan. 2003–mid-June 2004

The use of suicide attackers for effective and precise execution of terrorist attacks has tactical and operational advantages. Suicide attackers serve as human smart bombs, guiding themselves to their target with optimal timing for causing more fatalities and destruction. From an operational standpoint, the use of suicide attackers makes it unnecessary to provide an escape route for the perpetrators, and helps keep the terrorist organization's activity compartmentalized.

Suicide attacks were inspired to a large degree by a combination of radical Islamic fundamentalist philosophy and local nationalist motives. Al-Qaeda and its affiliates have had a great impact though by no means a monopoly on the spread of suicide terrorism (figure 3.2), and about twenty women, of Chechnyan, Uzbek, or Palestinian origin, also figured among the executors of the attacks. The addition of al-Qaeda and its affiliates to the circle of organizations using suicide terrorists, however, has imbued the phenomenon of suicide terrorism with a significant international dimension. The activity assumes al-Qaeda's global philosophy and suits its goal to have this specific form of operation proliferate – in view of its tactical advantages and symbolic significance – and become a keystone in the conflict between Islam and the rest of the world.

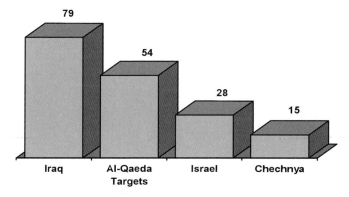

Figure 3.2 Distribution of Suicide Terrorists, Jan. 2003–mid-June 2004

The symbolic importance of the suicide attack weapon for al-Qaeda and its allies was equal to or greater than its operational advantages. For Osama Bin Laden, the sacrifice of life by al-Qaeda loyalists constitutes an extremely powerful propaganda tool for demonstrating the uncompromising devotion of his organization and protégés and their adherence to the way of God (*fi sabil Allah*). In Bin Laden's philosophy, suicide epitomizes the supreme level of faith in God, and suicide on behalf of God represents the moral superiority of the Muslim fighter over his infidel enemies. In his preaching, Bin Laden repeats the slogan, "We love death more than our enemies love life."[3] This slogan is designed to highlight the depth of Muslim belief, in contrast to the spiritual weakness, hedonism, and lack of values of their enemies. Muslims believe that the superiority of their faith guarantees ultimate victory over the infidel.

Al-Qaeda has made suicide attacks its trademark and has disseminated this philosophy among the organizations and terrorist networks that it supports. It boasted a large pool of volunteers from the various generations of "Afghan alumni" – veterans of the war in Afghanistan against the former Soviet Union, as well as second and third generation descendants, who arrived in Afghanistan at the beginning of the 1990s or later, underwent training and indoctrination, and swore allegiance to Bin Laden. Many expressed readiness to commit suicide for Islam and swore a personal oath of allegiance to the leader, all of which was recorded on video. This created an extensive cadre of potential suicide terrorists, who were later called on to fulfill their vows by conducting suicide attacks.[4]

From 2003 until mid-2004, al-Qaeda and its affiliates carried out suicide attacks in locations where, for the most part, they had not previously operated. After the military operation in Iraq officially ended, suicide attacks with multiple participants were staged and portrayed as retaliation for the targeted countries' support for the war. These attacks took place in Saudi Arabia in several waves and in Morocco. Other operations included an attempted combined suicide attack – ultimately thwarted – to cause mass casualties in Jordan. Terrorist elements supported by al-Qaeda also used suicide attackers in Uzbekistan and Pakistan. In April 2004, the Uzbek Islamic Movement, closely allied with al-Qaeda, carried out its first suicide terrorist attacks, striking against government targets in the capital of Uzbekistan. Dozens of people were injured in two separate terrorist assaults. In Pakistan, a terrorist attack was carried out against Pakistan President Pervez Musharraf whom Ayman al-Zawahiri, Bin Laden's deputy, labeled a traitor and collaborator with the enemies of Islam. Zawahiri called on army officers to rebel against Musharraf and kill him.[5] In addition, a number of suicide attacks were carried out in 2004 in Pakistan against Shiite targets, killing about seventy people.

➤ THE LOW INTENSITY CONFLICT IN IRAQ

The military campaign in Iraq embodied the use of military power advocated by the Bush administration in the wake of the September 11 attacks. Based on the "Bush doctrine," the war was designed to provide a deterrent model for other rogue regimes, and contribute to the war against international terrorism. Yet despite the aspirations and calculations of the US administration, most countries around the world did not seem to believe that the war in Iraq represented a primary effort in the war against terrorism.[6] Moreover, more than a year after the impressive military victory, it appears that while the architects of the war in Iraq hoped it would lend momentum to the international struggle against terrorism, the war has actually become an obstacle in this struggle, at least in the short term. The low intensity conflict that has followed the military campaign in Iraq has caused a large number of casualties among US military forces, the coalition allies, and Iraqi civilians. It has commanded much effort from the US administration, both in time spent by political and military leaders and in human resources and money. Furthermore, it has created momentum for many terrorist elements, but chiefly al-Qaeda and its affiliates. Ironically, the war cast in part as an integral fight against international terrorism prompted the chief targets of that war to exploit the situation in a beleaguered Iraq to reinvigorate their cadres and recruit new volunteers for what they describe as a defensive war against foreign occupation, or in terms of the global jihad, as a war to free "occupied Muslim soil."

The low intensity campaign in Iraq conferred many advantages on small and irregular forces, which exploited the prevailing chaotic situation to attack the numerous civilian and military international forces present, as well as symbols of the current Iraqi establishment. The widespread violence that became routine has hampered the formation of a stable regime to replace the disbanded Ba'ath regime, the reintroduction of public order, and the transfer of authority to an elected Iraqi government. Moreover, the post-war conflict enabled radical Islamic elements to use Iraq as an alternative theater for rehabilitating the image of al-Qaeda and its Taliban patrons, whose prestige was severely damaged by their crushing military defeat at the hands of the US-led coalition against terrorism. Al-Qaeda and its affiliates exploited the domestic unrest to strengthen their global jihad, recruit new cadres, and escalate their suicide attacks. Indeed, Iraq became their theater of choice for demonstrating their dedication to this type of operation. Drawing on their new activity in Iraq, these elements tried to strengthen their image as "winners," export suicide attacks to new areas, and make them a model for imitation by organizations around the world that have not yet employed this brand of terrorism.

The guerilla and terrorist actions in Iraq were undertaken by various elements: Shiites, Sunni supporters of the old regime, and Islamic fundamentalists linked to al-Qaeda and its affiliates. This non-allied source of insurgency included many who were not involved in subversive activities under the Saddam regime, which had ruthlessly suppressed any attempted uprising. For example, Shiite elements, led apparently by Muqtada al-Sadr, who challenged the more moderate Ali Sistani for the leadership of the Shiite majority in Iraq, were violently opposing American forces, primarily in the Shiite cities of Najaf and Karbala. Sadr even threatened to carry out suicide attacks against US military forces if they entered the holy cities. The ability of the pragmatic Shiite leaders to maintain relative restraint among the Shiites was to a large degree due less to their objections to terrorism per se than to their expectations that the Shiites would be able to utilize their numerical advantage to win a major share of any future administration in Iraq. Fighting as well were Sunni activists identified with Saddam Hussein's administration whom the media labeled "Saddam's fedayeen." These elements have been involved in guerilla and terrorist operations against the foreign forces and the Iraqi police as acts of revenge, in protest against their removal from positions of political power and with the aim of winning positions of power in any future government.

In addition, various other parties, those identified with al-Qaeda, have also taken an active part in guerilla and terrorist operations, including suicide attacks. Among these were the fundamentalist Kurdish Muslim organizations Ansar al-Islam and Ansar al-Sunna and foreign affiliates directed by members of the network run by Abu Musab al-Zarqawi. A Jordanian, Zarqawi was one of the most wanted men in the world and headed several al-Qaeda-supported terrorist networks active in the Middle East, Europe, and the Caucasus. A reflection of Zarqawi's key role in terrorist attacks in Iraq is a letter to senior al-Qaeda personnel that the Americans captured from one of his deputies in early 2004. In the letter, Zarqawi expressed concern that the battle for global jihad was losing ground in Iraq. His response was to recommend terrorist attacks against Shiite centers. He called the Shiites traitors to Islam, and advocated a "strategy of friction," i.e., terrorist attacks in Shiite areas, in order to incite a war between them and the other ethnic groups. He hoped that these attacks would lead to an ethnic civil war and a general destabilization of Iraq.[7] On another occasion, Zarqawi took direct public responsibility for a few notorious suicide attacks in Iraq.[8] Thus, while it was still difficult to identify exactly who was behind many of the guerrilla and terrorist attacks in Iraq, it appeared that at least the suicide attacks were undertaken largely by elements identified with al-Qaeda.

For al-Qaeda and its affiliates, Iraq has been the main stage on which they could use suicide operations to clearly demonstrate that they do not fear dying in defense of Islam. Indeed, as part of the psychological warfare waged by al-Qaeda, Bin Laden and Zawahiri called on Iraqis even before the war erupted to prepare to defend their country. They stressed the importance of suicide missions as a principal recourse for defeating their enemies,[9] and in fact, in the campaign following the war, suicide attacks have become a key tool for demonstrating opposition to the foreign forces in Iraq. They have also been an effective measure for striking at local law enforcement and military targets and at groups involved in the domestic conflict between the various ethnic groups (figures 3.3 and 3.4). Most of the perpetrators probably belonged to Sunni Islamic elements allied with al-Qaeda.

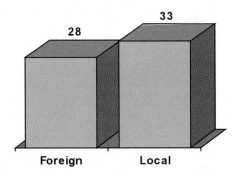

Figure 3.3   Suicide Terrorism Targets in Iraq, Jan. 2003–mid-June 2004

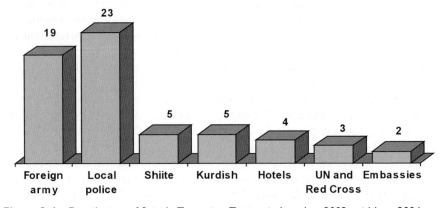

Figure 3.4   Distribution of Suicide Terrorism Targets in Iraq, Jan. 2003–mid-June 2004

Violence in Iraq, including the kidnapping of foreign media personnel and aid workers as hostages, was also committed by local gangs without any defined and concrete political demand. Interestingly, many of the initial kidnappings of foreigners in Iraq ended with the hostages released unharmed. Some subsequent events proved more violent, with hostages executed and with indications of links to al-Qaeda-affiliated groups. In both sets of cases, however, kidnappers demanded that the hostages' governments change their policy in Iraq, but conducted no serious negotiations with Western governments intended at exacting any concrete political concessions. This suggests that at least many of the kidnappings were random, perpetrated mostly for revenge, protest and propaganda, and purposes intended to demonstrate defiance against the occupiers rather than to extract any serious political concessions.

➤ AL-QAEDA AND ITS AFFILIATES BEYOND IRAQ

➤ Arab States

In 2003–2004, al-Qaeda and its affiliates have begun focusing on terrorist attacks in Arab countries considered "traitorous" for their cooperation with the West in general and the US in particular. The strikes were aimed at destabilizing these regimes, demonstrating their vulnerability through direct attacks on symbols of authority and mass killings of foreigners residing in those countries. The high profile terrorist attacks in Saudi Arabia were particularly prominent in this context, as part of a strategy of direct confrontation with the local security forces.

In Riyadh on May 12, 2003, a group of between nine and twelve al-Qaeda operatives divided into three squads and carried out coordinated suicide attacks against a residential complex for foreigners in the eastern part of the city. Thirty-one people were killed, among them foreigners, and dozens were injured. In November 2003, also in Riyadh, an al-Qaeda squad carried out another suicide attack against a residential complex for Arab foreign residents, killing seventeen people. This attack, in which primarily Muslims were killed, led to severe criticism of al-Qaeda, but al-Qaeda spokespersons claimed that those injured were agents and collaborators helping the FBI to interrogate al-Qaeda members. April 2004 witnessed another suicide attack in the Saudi Arabian capital, this time against an intelligence headquarters, which ended with four fatalities. The same month trucks loaded with bombs were seized in Saudi Arabia, one of which carried 1,200 kilograms of explosives designated for mass terrorist attacks in the country.[10]

The next two attacks were aimed against the oil industry of Saudi Arabia: on May 1, 2004 al-Qaeda attacked an oil company in Yanboa killing five people, and on May 29, a squad of four people attacked three compounds of foreign oil companies located in Khobar, killing twenty-two people. Most of the fatalities were foreigners who were selected for execution because of their religious (i.e., non-Muslim) affiliation. This attack, which was designed as a suicide operation, turned out to be a barricade hostage situation that ended with the escape of three of the four perpetrators.[11] Al-Qaeda's attacks led the Saudi Arabian regime to harsh countermeasures against the wave of terrorism. Apparently the Saudi Arabian rulers realized that elements formerly tolerated if not encouraged have turned against their host country.

Another Arab country that experienced terrorist attacks by al-Qaeda affiliates was Morocco. On May 16, 2003, thirteen suicide terrorists belonging to a local network with links to al-Qaeda divided into four squads and attacked Jewish and Western targets in Casablanca. These attacks killed thirty-one people and injured about sixty. Following the attacks, many arrests were made and further terrorist attacks in other Moroccan cities, including a number of suicide attacks, were thwarted. These arrests revealed an extensive infrastructure of radical Islamic operatives who acted with the help of elements from outside Morocco to realize the vision of a establishing an Islamic theocracy.

A mega-terrorist attack against three high profile targets was prevented in Jordan in April 2004. The targets included general intelligence headquarters in Jordan, the US embassy in Amman, and the prime minister's office. The attack was interrupted in the last stages of preparation following the arrest of members of the terrorist network, which consisted of Jordanians and Syrians acting on the instructions of Zarqawi. Jordan arrested a number of the prominent people planning the operation, including the commander, Azami Jiusi, a Jordanian of Palestinian origin, who had become an expert in explosives at the al-Qaeda camp in Afghanistan. In his interrogation, it was learned that he had operated under orders from Zarqawi. A few of the terrorists were killed during the arrests. Following the arrests, three trucks loaded with explosives that were ready to be detonated were seized, and warehouses where the group concealed their chemical materials were exposed. Jordanian explosives experts said that the planners of the attack had assembled twenty tons of explosives and seventy-one types of chemical agents, including a chemical causing third-degree burns, nerve gas, and chemicals causing asphyxiation.[12]

➤ Europe and Beyond

Europe as a field of action and a target for terrorist operations by al-Qaeda and the terrorist networks supported by it is not new. Consequently, the dramatic terrorist attack in Spain in March 2004 essentially came as no great surprise. Ideologically, al-Qaeda and its affiliates have always regarded Europe as one of the important theaters for global jihad, on the way to realizing their vision of an Islamic caliphate ruling the world. Europe is considered part of Dar al Harb, i.e., a region controlled by non-Muslims, which must be liberated in a war. Operationally, Europe meets a number of fundamental conditions that make it easy for al-Qaeda to operate in, including:

- Europe includes large Muslim minorities in various countries, which are reservoirs for new recruits. Most recruits are children of immigrants, born and raised in Europe yet who feel discrimination because of their ethnic origins. They resent their status as children of immigrants and feel alienated from European society.
- The open and liberal democratic lifestyle in Europe facilitates the free movement and passage of activists, equipment, and material, both within Europe and from Europe to other target countries.
- Local Islamic institutions such as mosques, charity organizations, and cultural centers provide infrastructures to recruit new agents, transfer funds, and coordinate secret activities.
- There is a pool of young people who have fallen into a life of crime and have been sent to prison. These people can be converted to Islam and recruited to the ranks of global jihad. While the potential for this exists elsewhere as well, the phenomenon is particularly pronounced in Europe.

The main function of al-Qaeda's infrastructure in European countries such as Germany, the UK, and Spain was long-term planning in preparation for the September 11 terrorist attacks. The subsequent exposure of these networks linked to al-Qaeda illustrated the high potential risk they pose to the security of Europe. Even earlier, however, starting in 2000, a number of terrorist attacks were foiled by the arrest of operatives in Germany, France, Italy, the UK, Belgium, and the Netherlands. Some of the terrorist attacks prevented involved non-conventional materials and suicide terrorists.[13] The preventive measures taken by the security forces in Europe were clearly inadequate but nonetheless served to give a sense of security (admittedly illusory) to Europeans, who until 2003 had not personally experienced the horrors of terrorist outrages committed by al-Qaeda and its affiliates. The gap between perception of the threat and the threat itself was evident in the limited European actions against al-Qaeda, coupled with tolerance of countries that supported the organizations, even if only passively.[14] A turning point in this perception began emerging following the waves of mass terrorist attacks in Turkey and Spain.

On November 15, 2003, twin terrorist attacks were carried out against synagogues in Istanbul, killing twenty-three people and injuring 300. A few days later, on November 20, the British consulate and a British bank in Istanbul were attacked; thirty-eight people were killed in this attack and 500 injured. Most of the casualties in these attacks were Muslims and Turkish citizens. Suicide squads belonging to a Turkish terrorist network supported by al-Qaeda carried out the attacks; the heads of the network were Afghan alumni who had met Bin Laden and trained at al-Qaeda camps. They then returned to their country and planned the attack, which was financed by unidentified parties from outside the country.[15]

The most dramatic and deadly attack to date on European soil was carried out on March 11, 2004 against the Madrid railway network by a terrorist network of expatriate Moroccans who enjoyed al-Qaeda support. One hundred and ninety-one people were killed and hundreds were injured in this series of coordinated explosions. The bombs, timed to explode at short intervals, were operated by cellular telephones in order to cause the maximum number of casualties. Technical problems prevented a number of explosive devices from detonating, while others detonated "prematurely" because trains had reached the previous station after their scheduled arrival time, thereby preventing casualties from reaching the much larger number hoped for by the perpetrators of the attack. Following the attack, an apartment in Madrid that was the hideout of the heads of the terrorist network was discovered. A few of the group's members were hiding there, along with weapons for additional terrorist attacks. On April 3, during the siege of the apartment by the Spanish security forces, seven inside committed suicide by setting off their explosive materials.

In this context it is worth highlighting the modus operandi of al-Qaeda and its allies, whose psychological warfare is waged mostly after their mass terrorist attacks have been carried out. Their goal is to heighten the fear caused by their violent activity in order to magnify the powerful image of global jihad and to threaten additional terrorist activities if their dictates are not accepted. An example of this is the recording published by Bin Laden after the attacks in Spain, in which he ostensibly offered a *hudna* (temporary ceasefire) to European countries withdrawing their forces from Iraq.[16]

The attack in Spain sharpened the perception of danger on the part of security forces throughout Europe. In addition to coordination and strategic cooperation between the security forces, the new attitude led to the additional arrests of those suspected of planning terrorist operations (in the UK in March, in France in April, and in Turkey in May). The arrests revealed preparations by al-Qaeda and its affiliates for further terrorist attacks in Europe. Following the investigation of the terrorist network that carried out the terrorist attacks in Casablanca in May 2003, direct connections were traced between the network and a terrorist infrastructure in London. The connections between members of terrorist networks allied with al-Qaeda operating throughout the world lay at the center of the al-Qaeda phenomenon. It was this organizational connection that provided the terrorist networks with the extensive infrastructure and operational flexibility necessary for their sophisticated terrorist activity.

Agents of al-Qaeda and its affiliates have extended their activities beyond European borders as well. In October 2003, Australia extradited to France Willie Brigitte, who belonged to a local terrorist network in France and who helped set up the infrastructure for terrorist attacks in Australia. His arrest led to the arrest of operatives of the network in Australia, and probably prevented a terrorist attack there.[17] In a separate incident, Christian Ganczarski, a German citizen of Polish origin who converted to Islam, was arrested in Paris in June 2003. Ganczarski, one example of the large number of "white Moors" in the ranks of al-Qaeda, was a friend of Nizar Nawar. Ganczarski was the last man to speak with Nawar by telephone before the latter carried out the al-Qaeda suicide attack on a synagogue in Djerba, Tunisia in April 2002. Ganczarski was arrested together with Karim Mahdi, who planned to carry out terrorist attacks on Reunion Island, a French possession in the Indian Ocean.

➤ SUCCESSES AND CHALLENGES IN THE WAR AGAINST AL-QAEDA

Since the international offensive was launched in Afghanistan in October 2001, a worldwide search for terrorists has been conducted as part of a global strategy for fighting terrorism. This strategy includes intelligence, police, and diplomatic efforts, which in 2003 generated increased cooperation between security services around the world on both the bilateral and multilateral levels. The result has been prevention of many terrorist attacks and several important arrests. Indeed, during this period several wanted senior terrorists at the highest operational levels of al-Qaeda and its satellite organizations have been arrested, along with the arrest of hundreds of operatives from al-Qaeda and its global affiliates.

The most impressive achievement in this respect was the capture of Khaled Sheikh Mohammed in Pakistan in March 2003. Mohammed directed and coordinated the September 11 attacks in the US on behalf of al-Qaeda, and was thus responsible for the murder of over 3,000 people. He was also the mastermind of other particularly spectacular terrorist attacks that were either carried out or were plotted and thwarted after the September 11 attacks. For example, in a planned terrorist attack foiled in Singapore in December 2001, seven trucks loaded with explosives were to have been blown up at the US, British, and Israeli embassies, and at US companies. Mohammed sent a suicide terrorist wearing explosives-laden shoes to blow up an American Airlines plane in December 2001. He was also responsible for suicide terrorists who detonated a car bomb at a synagogue in Djerba, Tunisia in April 2002, the car bomb terrorist attack on Bali Island in October 2002, in which approximately 200 people were killed, and other attacks. His arrest was made possible by bilateral cooperation between the US and Pakistan. Given the previous alliance between Pakistan, al-Qaeda, and the Taliban, al-Qaeda's patron, the joint efforts in the pursuit of al-Qaeda personnel were particularly significant.

Other prominent arrests included two of al-Qaeda's most experienced operational commanders, apprehended later in 2003: Walid bin Attash and Abd al-Rahim al-Nashiri, who were among the planners of the 2000 attack on the destroyer *USS Cole*. Riduan Ismuddin, better known as Hambali, chief operations officer of al-Jama'a al-Islamiyya, a regional terrorist organization extremely active in Southeast Asia, was captured in August 2003. Hambali was Khaled Sheikh Mohammed's senior partner, the main liaison between al-Qaeda and al-Jama'a al-Islamiyya, and the principal coordinating and financing pipeline for the terrorist attack in Bali and a planned terrorist campaign in Asia.[18]

The pursuit of al-Qaeda extended to Africa, where the organization carried out terrorist attacks against an Arkia Airlines passenger jet and Israeli tourists in the Paradise Hotel in Kenya in November 2002, and was preparing additional spectacular terrorist attacks. These attacks were foiled with the arrests of several members of the local terrorist network, who were scheduled to stand trial in Kenya during 2004.[19] The unrelenting pursuit of other members of the network, probably in hiding in Kenya and Somalia, was also stepped up.

Despite these significant tactical successes, security services have not managed to capture al-Qaeda leaders Bin Laden and Zawahiri. This ongoing primary objective essentially amounts to a failure, and has partially obscured the achievements. The capture or killing of the two men is an essential condition, even if not a self-sufficient one, in the war against the global terrorism of al-Qaeda and its affiliates. In addition to constituting a symbol and an example to all global jihad operatives and supporters, Bin Laden has taken part in the actual planning and authorization of al-Qaeda's terrorist attacks. He has continued to lead the organization and to incite his followers to further their terrorist activities throughout the world. Zawahiri, who was responsible for al-Qaeda's non-conventional activity, remained the organization's chief ideologist. He has preached his venom against anyone considered an enemy of Islam in the video and audio recordings that he published regularly calling for continued jihad and the mass killing of infidels. At the same time, however, it was clear that due to the prolonged tolerance, however passive, of the Afghan alumni in the years before the September 11 attacks, even the disappearance of these two leaders in one way or another would neither eliminate al-Qaeda and its affiliates nor remove at a single blow the threat of fundamentalist Islamic terrorism from the international arena.

The pursuit of Bin Laden and Zawahiri was based on cooperative efforts of the US and Pakistan and the special forces they employed for military attacks and raids in the border area between Pakistan and Afghanistan. These attacks aimed at arresting the two terrorist leaders and striking at operatives of al-Qaeda and the Taliban. Local topographical and cultural conditions hampered these actions: ground conditions hindered movement, especially given the many caves and other hiding places, and the harsh weather restricted military operations during part of the year. In addition, the tribal structure in the region was highly decentralized, and many of the people in the area were sympathetic to the Taliban and al-Qaeda. In March 2004, hundreds of al-

Qaeda members and Taliban supporters were arrested following a series of fierce battles in the region with Pakistani troops who apparently operated with the aid of special American forces lying in ambush on the Afghanistan side of the border. Reports that Zawahiri had been captured in this campaign proved to be mistaken, but the joint efforts of Pakistan and the US to achieve a decisive victory over al-Qaeda and the Taliban were expected to continue until the leadership was captured.

Achievements in the war against terrorism were accompanied by the contribution of countries that had formerly assisted terrorism to a more limited degree, particularly against al-Qaeda and its affiliates. Saudi Arabia, for example, in the aftermath of its having been an al-Qaeda target, has demonstrated new cooperation. Sudan likewise has contributed to the effort, as has Libya, in the wake of its gradual disassociation from international terrorism. In Africa, which has proven a fertile field for activity by al-Qaeda and its affiliates, various countries, first and foremost Kenya, have also shown increased willingness to participate in the joint effort against the terrorist networks operating in the region. Kenya, the site of two al-Qaeda attacks, joined the US-East Africa Counter-Terrorism Initiative (EACTI), to bolster its ability to meet the challenge of terrorism on the operative, governmental, and legal planes.[20]

As for the principal challenges facing the countries in the war against international terrorism in 2004, to a large extent these remained similar to those of the previous decade. The September 11 attacks and ensuing developments, however, have made the need to deal with these challenges more urgent and acute. The main challenge was to act in coordinated fashion, systematically, and intensively to neutralize the terrorist networks of al-Qaeda and its allies. Leading objectives were that the top leaders be captured or killed, the clandestine terrorist networks be exposed, and their attempts to acquire, manufacture, and use non-conventional weapons be obstructed. The policy of countries actively supporting terrorism and that of countries granting passive aid to terrorists on their soil was marked for transformation. Indeed, the September 11 attacks highlighted the problem of the "failed states," countries that have refrained from directly confronting the terrorists residing on their soil, due either to a lack of political will on their part to confront population groups providing refuge to terrorists, or to their inability to act effectively against the terrorists among them.[21]

In Iraq, the violent struggle of the forces acting to stabilize the country and establish a temporary government in preparation for elections and the creation of a long-term government versus the forces trying to perpetuate anarchy was expected to continue. The goal of the latter was to undermine the achievements of the US and force it into a shameful withdrawal from Iraq. The result, therefore, may be a protracted low level conflict in Iraq, particularly if efforts fail to establish an effective Iraqi government with broad, stable, and effective support. In that case, the situation is liable to escalate, and Iraq could be expected to continue being a focus for domestic international terrorist activity against foreign forces present in the country.

Al-Qaeda and its affiliates can also be expected to continue perpetrating terrorism in the international arena, while using events in Iraq to justify the actions they are planning in any event. The emerging trend in activity by the terrorist networks of al-Qaeda and its allies against Arab regimes defined as collaborators was likely to continue and perhaps even spread to other countries in the Persian Gulf and the Middle East, in accordance with the campaign of vituperation published on internet sites and on recordings in the name of global jihad.

The past year has furnished Europe with sufficient hard and painful evidence that it too figures among the targets of al-Qaeda and its affiliates, with or without a direct connection to their Islamic agenda, since al-Qaeda regards it as an integral part of its declared enemies. As such, European countries were liable to suffer further terrorist attacks by terrorist networks in their country supported by al-Qaeda when operational readiness for these attacks would "mature," and not necessarily according to political circumstances. Bin Laden and his men were expected to continue trying to expand the circle of suicide terrorists and to attempt attacking in countries where they have not yet operated.

One of the key challenges that was expected to continue occupying defense agencies throughout the world was preventing the realization of threats by al-Qaeda spokesmen as publicized on the internet, particularly toward the end of the 2003. These spokesmen affirmed that al-Qaeda was planning a mega-terrorist attack of such dimensions as to completely overshadow the September 11 attacks.[22] The question posed by these threats was whether they reflected true capabilities and genuine intentions, or whether they were part of the psychological warfare that the organization has been conducting successfully for several years. The drive of al-Qaeda and affiliated terrorist cells to carry out attacks using non-conventional materials also remained to be foiled.

The potential threat of terrorist activity together with the heavy costs in human life and physical damage when those threats were translated into terrorist activity likewise exacted a high economic price. Severe economic damage occurred in 2003 in civil aviation and tourism, as a result of the cancellation of many flights in Europe, the US, and Africa.[23] Countries hitherto deemed safe for tourism have been marked for periods of time as not recommended for tourist visits due to warnings of impending terrorist attacks. Thus, the substantial direct and indirect economic burden generated by the threat of terrorism and terrorist activities commanded serious attention among leaders deciding whether or not to join the battle against terrorism. Clearly, the criterion of whether or not a country had actually been attacked so far was of marginal value, if any, since terrorism has shown to strike at what might otherwise be presumed the most unlikely of targets.

The second half of 2004 was scheduled to include a number of events with a particularly high media profile, and these were liable to constitute an attractive target for terrorist organizations. The most prominent was the Olympic Games in Greece: the high media profile and symbolic significance of the Olympics prompted much advance security work to thwart any attempts by al-Qaeda and its affiliates. Moreover, al-Qaeda is known to have intended an attack at the Sydney Olympics, but the attack was never actually staged due to an internal dispute in the terrorist cell in Australia entrusted with the mission. That the Olympics were already a target four years ago suggests that they might be a likely focus for al-Qaeda in 2004 as well. The presidential elections in the US were also liable to motivate al-Qaeda, which might try to redouble efforts to carry out a dramatic terrorist attack in order to challenge the leadership of President Bush and demonstrate the inadequacy of the campaign against them as well as its leader.

Almost three years after the September 11 attacks, it was evident that the struggle against international terrorism was likely to continue for a long time. The challenges facing the leading countries in the free world in this struggle were intricate and complex, due to the extreme and uncompromising character of al-Qaeda and its affiliates, who constituted a stubborn and dangerous enemy. Nor has the broad-based and well-coordinated coalition thus far forced countries that support terrorism to pay a high price for their policy in this area. As long as this situation continues, the chances of reducing the extent and risks of international terrorism will be low, and the threat posed by international terrorism to the peace and personal security of the people of the world will escalate.

Notes

1 See, for example, Michael Evans, "MI5 Chief Fears Dirty Bomb is Inevitable," *The Times*, June 18, 2003.
2 Suicide attacks began in Chechnya and Russia in 2000. Sixty-seven terrorists committed twenty-one suicide attacks in which 465 people were killed since 2000.
3 "As Taliban Collapses, Hunting al-Qaeda and Bin Laden Becomes Top Priority," *The White House Bulletin*, November 19, 2001.
4 For example, the recorded wills of the September 11, 2001 suicide terrorists, and the recorded wills of the perpetrators of the August 8, 2003 attacks in Riyadh.
5 "Purported Zawahiri Tape Condemns Musharraf; No. 2 Figure in al-Qaeda Is Said to Call for Overthrow of Pakistani Government," *Washington Post*, March 26, 2004.
6 Yoram Schweitzer, "The War in Iraq and International Terrorism," in Shai Feldman, ed., *After the War in Iraq: Defining the New Strategic Balance*. Brighton & Portland: Sussex Academic Press, 2003, pp. 45–54.
7 Douglas Jehl, "U.S. Aides Report Evidence Tying al-Qaeda to Attacks," *New York Times*, February 10, 2004.
8 Schmitt, "Allies Suspect al-Qaeda Link to Bombings in Basra; Death Toll Is Reduced," *New York Times*, April 23, 2004.
9 "Threats and Responses: The Terrorists," *New York Times*, February 12, 2003.
10 Zvi Barel, "Now They're Not Donating Camels Anymore," *Ha'aretz*, May 7, 2004, B5.
11 "Commander of the Khobar Terrorist Squad Tells the Story of the Operation," MEMRI, No.731, June 15, 2004.
12 "Jordan Says Major al-Qaeda Plot Disrupted," <www.cnn.com>, April 26, 2004.
13 Yoram Schweitzer, "The Age of Non-Conventional Terrorism," *Strategic Assessment* 6, no. 1 (2003): 26–31.
14 Jonathan Stevenson, "How Europe and America Defend Themselves," *Foreign Affairs* 82, no. 2, March–April 2003.
15 Yossi Melman, "Damascus Extradites 22 Suspected of Involvement in Terrorist Attacks to Turkey," *Ha'aretz*, December 13, 2003.
16 "Osama Bin Laden Speech Offers Peace Treaty with Europe, Says al-Qaida 'Will Persist in Fighting the U.S,'" *MEMRI*, April 15, 2004.
17 Martin Chulov, "Terror in Our Midst," *The Australian*, March 23, 2004.
18 Yoram Schweitzer, "The Capture of Hambali: Glass Half Empty, Glass Half Full," *Tel Aviv Notes* no. 84, August 17, 2003.
19 Matthew Rosenberg, "Kenyan Tells of Plot to Bomb Embassy," *Washington Post*, October 25, 2003.
20 "Africa Overview," *Patterns of Global Terrorism*, U.S. Department of State, April 29, 2004, pp 1–3.
21 Yoram Schweitzer, "Neutralizing Terrorism-Sponsoring States: The Libyan 'Model,'" *Strategic Assessment* 7, no. 1 (2004): 7–14.
22 The principal website used to disseminate these messages is <http://www.cybcity.com>.
23 Sara Goo Kehaulani and Dan Eggen, "Security Threat Lowered to Yellow," *Washington Post*, January 11, 2004.

4

WMD Proliferation Trends and Strategies of Control

EMILY B. LANDAU AND RAM EREZ

When we compare current assessments of weapons of mass destruction (WMD) proliferation in the Middle East with common estimates from just a few years ago, what is most striking is how very different the picture looks today. In fact, however, the change has not occurred in the reality itself, but rather in our knowledge of that reality. The amount of information that came to light in 2002–2004 about WMD advances in Iran and Libya, and especially in the nuclear realm, is unprecedented. It was revealed that nuclear activities in both states were ongoing for close to two decades.

Attention centered initially on nuclear activities in Iran, following critical disclosures by an Iranian opposition group in the summer of 2002. Libya's willingness to submit all its WMD facilities to international inspection, announced in December 2003, broadened the focus to include that country as well. In January 2004, following investigations in Libya, new information surfaced about the global reach of the Pakistani "nuclear network." With the spotlight on the extensive and intricate black market in nuclear technology, parts, and plans, speculations increased as to the activities in additional Middle Eastern states, including Syria, Saudi Arabia, and even Egypt. Overall, new revelations on nuclear-related activities abounded in this period.

In direct contrast was the surprise of Iraq, a major focus of attention in the WMD realm from the 1990s onward. Though not final, in 2004 the emerging picture is one of gross *over*estimation as to the actual WMD threat Iraq had posed on the eve of the war. The post-war findings – or lack thereof – challenge the common assumptions regarding the WMD arsenals and development programs that had circulated previously.

Rather than discussing the possible intelligence failures latent in these incomplete WMD assessments, this chapter will dwell on the implications of the newly revealed cases of proliferation for international arms control and non-proliferation efforts. Thus, the focus is directed to those cases where there are strong indications that proliferation has occurred and to the steps that have been taken to confront the suspected proliferators. The analysis below opens by establishing a context for recent developments in proliferation, and then proceeds to a state-by-state review of Iran, Libya, Syria, Saudi Arabia, and Egypt on the basis of the information that came to light in 2002–2004. A brief look at Israel closes the list. Finally, reflections on Iran, the role of strong international actors in arms control efforts, and the need to address state motivation and security concerns will command special attention in the concluding remarks.

➤ NEW PROLIFERATION CONCERNS

Recent discoveries have channeled attention first of all to the potential threat implied by the proliferation that has taken place in the region. Of equal concern, however, are the troublesome questions these discoveries raised as to the ability of the International Atomic Energy Agency (IAEA) to detect signs of problematic nuclear activity as well as to the overall effectiveness of the non-proliferation regime.

The sense of crisis that these discoveries signaled, however, was mitigated by other factors, such as the Libyan decision to "come clean" in the context of a deal worked out with Britain and the US. The fact that a state like Libya, which had worked toward creating a non-conventional capability, made a decision to eliminate its WMD programs and allow unfettered inspections of its facilities was highly significant in terms of the message of willing cooperation that it imparted. Moreover, not only did Libya decide to disclose the details of its own WMD program, but by submitting its facilities to inspection, additional important information was revealed regarding the clandestine nuclear network run by Pakistan's top nuclear scientist, Dr. Abdul Qadeer Khan. This in turn led to further discoveries regarding other Middle Eastern states. Thus while United States determination to use military force in Iraq seems to have been at least a contributing factor to Muammar Qaddafi's decision to work with the US rather than against it, the Libyan case underscores that a cooperative relationship can be created.

Furthermore, as proliferation trends have become known, new modes of confronting these developments have begun to emerge in response. Thus, while newly revealed cases of proliferation have exposed the serious limitations of current international arms control treaties to stem proliferation, in the wake of September 11 and the Iraq War the ground has concurrently ripened for strong international actors to take the initiative and carve out a role for themselves in dealing with proliferation challenges. This has also helped to alleviate the sense that proliferation was spiraling out of control.

The Iraq War was a watershed in this regard, especially as far as US-European coordination of purpose. The war evoked crisis sentiments in the seemingly contradictory trends that it brought to the surface in how the US and Europe dealt with WMD threats. However, the overt trans-Atlantic dispute was in fact more over the validity of the threat assessments themselves than about the means for dealing with the threat. More importantly, when it became clear following the war that the US had not simply decided to replace diplomacy with military force on all fronts, the political climate began to change. Diplomatic efforts appeared that underscored the concern and commitment of both the US and key European states to confront the challenge of WMD proliferation, including a new ad hoc multilateral initiative – the Proliferation

Security Initiative (PSI) – that was quickly agreed upon by a number of states as a means for dealing with new proliferation suspicions and realities.[1] There has been a growing sense that new modes of coordinated action are needed, in order to complement and enhance the existing arms control regime.

Interestingly enough, the failure to find WMD stockpiles in Iraq did not detract from the increasing anxiety as to the suspected situation in Iran or Syria. On the contrary, in confronting these states, there was a significant degree of convergence between the US and key European states not only on the assessments themselves, but also on the need for adopting a tougher stance in challenging suspected proliferators. This emerged, for example, in the position adopted by the European Union (EU) in its bilateral relations with Syria – linking the signing of an Association Agreement with Syria to clear steps that Syria must take in order to reverse its presumed proliferation efforts.

Because political and diplomatic efforts are still in a state of flux, however, the political will of the US and European states to take action has not yet resulted in fully developed or wholly consistent arms control channels. There has been evidence of hesitation and certain conflicting tendencies, most notably regarding US attitudes toward Pakistan since the nuclear network revelations. There remains no doubt about Pakistan's role in promoting global proliferation, but the US has had little interest in cracking down on this state due to its cooperation in the war on terrorism.

Moreover, as the analysis of developments in Iran will argue, there were cases where the new determination may have come too late to reverse the direction of nuclear weapons development. This highlights a major limitation of the existing non-proliferation regime, namely, that the NPT is not designed to actively stop a determined proliferator from developing a breakout nuclear capability.[2] Managing the situation in Iran may ultimately require new regional arms control efforts that address broader security concerns, along the lines of what is pursued in relation to North Korea. Overall, it remains to be seen how policies will consolidate and catch up to the pace of events, and how new directions will integrate with the existing non-proliferation regime.

One of the most troubling aspects of the recent revelations concerns the activities of the nuclear black market – in particular, the range of international contacts that A. Q. Khan, the so-called "father" of the Pakistani nuclear project, established in the early 1980s and kept active until late 2003.[3] Khan's nuclear network operated as kind of a "supermarket" for states that tried to promote their nuclear program outside the international non-proliferation regime. Numerous companies around the globe, including German, South African, Japanese, and Malaysian firms, participated in the network by supplying various forms of equipment. Khan also operated through middlemen in Europe, Sri Lanka, Malaysia, and the UAE who facilitated the transactions.[4]

The Pakistani nuclear black market was widely exposed in the international media following Iran's report to the IAEA in November 2003 where it linked its supplies to Pakistan, and even more so following Libya's revelations a month later. However, as noted by CIA director George Tenet in February 2004, intelligence services were aware that illicit trade in nuclear parts and technology had been conducted with Pakistani orchestration for years.[5] When approached by the US with this information in late 2003, Pakistani President Pervez Musharraf claimed no knowledge of these activities. Pressure on Pakistan resulted in a public confession by Khan, claiming that he acted

61

without the knowledge of the Pakistani government. He was thereafter pardoned by Musharraf in February 2004.

Much is still unknown about the activities of the Pakistani network, and as IAEA Director General ElBaradei claimed in early 2004, what was known at that time was only the "tip of an iceberg" of the nuclear black market.[6] The fact that it became clear that nuclear components, materials, and designs can be (and have been) acquired "ready made" raised serious implications for estimates as to when a particular country could be expected to reach nuclear weapons capability. The drastically reduced time period demanded for this capability has given added impetus to the need for determined and coordinated international action to deal with proliferation attempts.

➤ STATE-BY-STATE REVIEW

➤ Iran

In August 2002, the National Council of Resistance of Iran, an exiled Iranian opposition group, announced at a news conference in Washington that it had evidence that Iran was pursuing clandestine nuclear activities, unknown to the IAEA, including top secret projects south of Tehran (figure 4.1). These consisted of a suspected uranium enrichment plant near Natanz and a suspected facility for producing heavy water in Arak.[7] This disclosure marked a turning point in the US anxiety over Iran's nuclear activity. Until then, Iran had been a source of concern, but the focus of attention had been Russian-Iranian cooperation on the construction of the Bushehr nuclear reactor. The US wanted to persuade Russia to take steps to ensure that Iran could not divert the civilian activity there to weapons-related activities. In the wake of the 2002 announcement, attention shifted from Bushehr to the undeclared sites, and European states and the IAEA became much more involved as well. The year 2003 was marked by intensive efforts on the part of the IAEA to comprehend the full scope and nature of nuclear-related activities in Iran.

Disclosures on nuclear activity in Iran that began with the National Council claim led to a major shift in estimations of the situation in Iran, and following the initial revelations, Iran's nuclear activities came under much closer international scrutiny. As a party to the NPT, Iran legally promotes a civilian nuclear program under the supervision of the IAEA. However, the new information that emerged suggested that Iran has promoted a parallel covert and comprehensive nuclear program with military significance. The Iranian argument that all new discoveries were part of the state's peaceful civilian program, framed by the oft-mentioned Iranian contention that nuclear weapons contradict Islam,[8] became increasingly difficult to accept. The US, key European states, and the IAEA have all undertaken an active role in pressuring Iran to comply with and expand its obligations according to the NPT. Iran, for its part, has so far followed a course of limited cooperation, with a pattern of one step forward and half a step back. In its reports, the IAEA, though highly critical of Iran's past behavior, has preferred to focus more on evidence of Iranian cooperation with the international agency, and has stopped short of determining that Iran had a nuclear weapons program in clear violation of its NPT obligations. However, IAEA patience with Iran is wearing thin, as evidenced by its June 2004 resolution.[9]

Figure 4.1 Nuclear Sites in Iran

Following the initial revelations, ElBaradei visited the Natanz site in February 2003 and found a comprehensive and sophisticated centrifuge uranium enrichment facility. In his June 2003 report to the IAEA Board of Governors he concluded that "Iran has failed to meet its obligations under its Safeguards Agreement."[10] In September, the Board issued an ultimatum to Iran calling upon it to cooperate fully with the IAEA, to suspend all further uranium enrichment-related activities, and to sign and implement the Additional Protocol by the end of October 2003.[11] In October, after close to a year of pressure and investigations into Iran's nuclear activities and ten days before the September ultimatum was due to expire, France, Germany, and Britain were able to work out an agreement with Iran, which was adopted by the IAEA Board of Governors in late November 2003. Iran admitted to having worked on secret nuclear plans for eighteen years and agreed to suspend all uranium enrichment-related activities as well as to join the Additional Protocol.[12] Iran in fact signed the Additional Protocol in December 2003, which allows for more intrusive international inspections, but six months later had yet to ratify it.[13]

Iran's network of suppliers was also better understood after Iran submitted a report on the history of its nuclear activities to the IAEA in November 2003. In addition to the known Russian, Chinese, and North Korean assistance, the investigators concluded that Pakistan had played a central role in supplying Iran with equipment and knowledge. This information began the public disclosure of the Pakistani supply network that had helped the efforts of Libya and North Korea as well.[14] Interestingly, when more evidence of the Pakistani proliferation network surfaced following the Libyan move, Iran denied receiving assistance from foreign scientists, arguing that the achievements were made by Iranian scientists and that only some individuals, acting as middlemen, had helped Iran buy equipment in the open market.[15] These claims were weakened by the admissions of the Pakistani scientists themselves.[16]

Other cracks in Iran's credibility appeared soon, in this case as to Iran's commitment to halt its nuclear ambitions. New evidence was introduced proving Iran was still concealing its nuclear efforts, including indications that Iran continued to assemble gas centrifuges and that it did not admit to holding designs of a more sophisticated centrifuge (P2), which could substantially increase its ability to enrich uranium.

In the wake of international pressure, Iran once again agreed to discontinue its centrifuge activities in late February 2004,[17] shortly before the IAEA Board of Governors was due to discuss developments in Iran. In March 2004, the Board published a resolution condemning Iran for failure to declare potentially arms-related activities, while commending Iran for its willingness to cooperate with the inspectors. Despite the new findings, the Board did not find Iran in breach of the NPT, which would have resulted in transfer of the issue to the UN Security Council.

Iran's behavior in March–April 2004 provided a concise illustration of its vacillating pattern between cooperation with the IAEA and attempts to advance its nuclear program. While the IAEA was discussing the wording of the March resolution, Iran made a tactical move that was interpreted by diplomats in Vienna as an attempt to signal to the Board that it might stop cooperating with the IAEA altogether if the resolution proved too critical. Under the pretext of a national holiday, Iran's ambassador to the IAEA announced the postponement of a planned inspectors' visit to Iran, from late March to late April.[18] Three days later, Iran backed down, and agreed to allow the scheduled visit.[19] Less than two weeks later, immediately following the inspectors' arrival in Iran, the director general of Iran's Atomic Energy Organization announced the renewal of Iran's uranium enrichment activities at its uranium conversion facility in Isfahan.[20] A few days later, and after harsh European criticism stating that Iran was sending "the wrong signal,"[21] Iran announced that it would not renew the program and vowed to speed up nuclear cooperation with the UN.[22]

Iran demonstrated that it was highly motivated to achieve a complete nuclear fuel cycle, and its limited, guarded, and conditional cooperation has not provided assurances that this ambition has changed. The question remains how far Iran is willing to go in confronting the international community on this issue. To date, Iran has played a clever game in balancing its cooperation with the IAEA, the US, and European states with efforts to pursue its strategic goal further. Iran has disclosed enough information both to demonstrate a positive attitude and to forestall the involvement of the Security Council, but at the same time it has persisted in concealing information that might enable a continued nuclear program. Iran walks a very thin line in this regard: it makes every effort not to violate the letter of the law, but its concerted ambiguity is gradually eroding any sense that it intends to uphold the spirit of the NPT.

The facility at Natanz is a case in point. Technically, Iran did not violate the NPT when it did not declare the centrifuges site near Natanz since, according to the NPT, a signatory state must report a new site within 180 days from the beginning of its operation and not in advance of operation. However, at that site the IAEA found a pilot plant of centrifuges for uranium enrichment and a commercial plant still under construction designed to hold 50,000 centrifuges. If completed, the plant is projected to be able to produce enough enriched uranium for twenty-five to thirty nuclear weapons per year.[23] The IAEA inspectors found traces of highly enriched uranium (HEU), which is of military grade, at Natanz as well as at the Kalaya Electric Company in an inspection of the summer of 2003. Iran did not admit to conducting tests with HEU, and claimed that these traces originated in contaminated imported parts of the centrifuges that had gone through several middlemen, making the source of the uranium impossible to trace.

A report of the National Council of Resistance of Iran from early 2004 claimed that developments of late 2003 indicated that not only had Iran not abandoned its nuclear aspirations; it had concluded that it must step up the pace of development. According to the report, Iran transferred its uranium enrichment activities to a series of smaller sites that remain undetected. Such activity would enable Iran to proceed more quickly on the road to nuclear weapons and to reach its first nuclear device as early as 2005.[24] For its part, Iran had voiced its expectation that all inspections would end by the June 2004 IAEA Board of Governors meeting, but this did not occur. ElBaradei issued a relatively harsh report on Iran's lack of cooperation, and demanded that Iran provide a full account within the next few months.[25] The Board of Governors resolution echoed ElBaradei's conclusions.

This review of developments in Iran, particularly the way Iran has been able to technically abide by the law while continuing with its nuclear program, suggests that as long as Iran continues to pursue the goal of mastering the complete nuclear fuel cycle under NPT supervision, it might have a way to beat the system. By the time the international community firmly addresses this shortcoming – if indeed it is possible to do so – Iran could pass the critical stage that would enable it to achieve its nuclear ambitions. This has made the situation ripe for strong international actors to assume a role in trying to fill the gaps inherent in the non-proliferation regime. Clearly, however, Iran has not facilitated these international efforts, and in fact has worked actively to challenge many of them.

Moreover, while the US has shown determination in acting on Iran, the 2002–2004 period has also seen the Bush administration in the process of reevaluating its position toward Iran. In December 2003, Powell was quoted as saying that the US was open to restoring a dialogue with Iran, and that it seemed to him that there is "a new attitude in Iran."[26] This was a substantial change in the American position since Iran's assignment to the "axis of evil" and even from the continued US attempts to have the Iranian issue referred to the UN Security Council. The US also presented a more moderate position with regard to the IAEA resolution in March 2004, when it refrained from insisting on referral of the Iranian issue to the Security Council. Nevertheless, although the US put greater emphasis on diplomacy, it remained highly critical of Iran.

United States determination to deal firmly with Iran encouraged the EU to take a more active stand on the threat as well. In fact, Iran's nuclear activity became an important issue for the EU as it strove to formulate a common foreign and security policy, and provided an opportunity after the Iraq War to play a more central role in containing WMD proliferation. The facts that came to light regarding the mature state of Iran's nuclear project made it easier for European states to agree that there was a real potential threat to their societies. As to dealing with the threat, their preference was to handle it diplomatically, using economic and political leverage. Based on these parameters, France, Germany, and Britain obtained an agreement with Iran in October 2003. Since Europe was Iran's principal trading partner and with a trade agreement between the EU and Iran under discussion, the EU was in a particularly strong position to influence Iran's strategic calculations.[27]

Overall, US determination not to allow an Iranian nuclear bomb, together with a firmer European stand on this issue, has the best chance of denying Iran a nuclear weapons option. However, even if the US and the Europeans are able to sustain long-term pressure, the question remains whether time is not playing to Iran's advantage. In other words, will Iran be able to reach a nuclear capability before it is found to be in clear breach of its obligations.

Complicating the efforts of the IAEA, the US, and European states to confront the Iranian challenge and to find it in violation of its commitments are the different interpretations at play. There was a question of interpretation of the evidence itself, of what Iran in fact agreed to in October 2003 (whether merely suspension of uranium enrichment, as Iran saw it, or cessation of these activities, as others interpreted it), and finally of what constituted a definite breach of NPT obligations. The problem of interpretation complicated coordination between the US and Europe, as well as the effort to send a clear message to Iran. It underscored that much of what was on the scene fell into the grey area of the kind of signals that Iran was perceived to be sending – whether of goodwill and a cooperative spirit or of attempts to beat the system. The Iranian approach seemed much more cooperative when compared with Iraq's record of dealing with international inspection agencies throughout the 1990s; however, when juxtaposed with the recent Libyan approach, it seemed qualitatively less so, especially as more and more information came to light.

Even were there an agreement that Iran was not upholding the spirit of the NPT, the question is whether action could be taken on this basis. In other words, under the present system, with the NPT as the basis for evaluating Iran's behavior, it could be that even with the new and coordinated determination of the US and Europe, Iran may be "beyond catching." For its part, even after the Iraq War, Iran might be looking at some other cases of proliferation, where once a nuclear capability was established, the international community ultimately had little choice but to live with it.

➤ Libya

While Libya has always been a thorn in the WMD field, particularly regarding its chemical weapons activities, comparatively less attention was focused on this state prior to the developments of late 2003.[28] It therefore came as quite a surprise when on December 19, 2003, Libyan leader Qaddafi announced that Libya would abandon all non-conventional weapons programs, including its nuclear, chemical, and biological projects. Qaddafi stated that he would open the country to international inspections, sign the Additional Protocol to the NPT, and join the Chemical Weapons Convention.[29]

The announcement came in the wake of nine months of secret negotiations with Libya conducted by Britain and the US. Libya approached Britain in March 2003 just before the Iraq War began, and the US joined the efforts shortly thereafter. These contacts included meetings in London and Tripoli, as well as visits of American and British experts to WMD facilities in Libya.[30]

In the weeks following Qaddafi's announcement, new information revealed the extent of Libya's WMD programs, which dramatically altered what was thought to be known on Libya's WMD programs and on the proliferation of WMD in the Middle East at large.[31] Although it was soon assessed that Libya was still years from reaching a nuclear capability, a critical factor was its ability to bypass international verification mechanisms as well as Western intelligence. Further information under review is expected to produce additional revelations, but initial findings have shed light on the Pakistani supply network and have implicated additional regional players, including Iran, Iraq, Saudi Arabia, Syria, and Egypt. Apparently, Libya had an off-the-shelf nuclear project that it bought from Pakistan. Libya did not possess the advanced scientific and technological infrastructure needed for the construction of an independent nuclear project, and had invested primarily in acquiring turnkey equipment.[32] Reportedly, Libya made a strategic decision in 1995 to increase its nuclear efforts, and was successful in obtaining gas centrifuges and uranium.[33] In addition to its nuclear program, Libya also maintained chemical and biological programs. According to a declaration of March 2004, Libya produced a mustard agent and had substantial quantities of material for the production of additional mustard and sarin agents.[34]

It is not yet clear what Qaddafi's motivation was in overturning his long-standing clandestine policy and acknowledging his WMD programs. Commonly held explanations assumed this was either the culmination of the long-term international sanctions on Libya; a reaction to the US determination to use military force against Iraq (with the added impact of the seizure of a shipment of centrifuge parts on its way to Libya in September 2003);[35] or Qaddafi's way of guaranteeing that his son would succeed him. Most likely, it was a combination of factors, with US determination providing a major impetus. In any case, Qaddafi's move set a precedent in several areas. First, it

67

showed that even a "rogue state" can change its policy: if proof was needed, Qaddafi demonstrated that the possibility of reasoning with such states does exist. Second, the fact that Libya, with its legacy of terrorism and clashes with the West, managed to reach an agreement with Britain and the US was significant. For other states with a history of bad relations with the US, this demonstrates that it is never too late to create a new basis for relations, and that such a change does not have to be preceded by an overthrow of the ruling regime.

Third, based on the Libyan precedent, the international community was on stronger ground when it demanded of other proliferating states that they renounce their WMD programs. Finally, the Libyan case set the precedent for negotiations and inspections of WMD proliferators outside the arms control regime. Following Libyan overtures, it was the joint action of Britain and the US that brought about the much hailed results. The IAEA was allowed to participate in the dismantling of Libya's nuclear program and was granted the responsibility of verifying Libya's future compliance with the NPT. However, it was the US and Britain that assumed the central role in the dismantling of Libya's WMD.[36] Although this might suggest that the global non-proliferation regime has been further weakened, at the same time it underscores that dominant powers can help fill the gaps that exist in the current regime with determined action, at least until these gaps are closed with new mechanisms.

➤ Syria

According to intelligence reports, in addition to stockpiles of ballistic missiles, Syria also has a well developed chemical weapons program. It is similarly estimated that Syria is active in the biological realm, although with a program smaller than in the chemical realm.[37] Regarding its nuclear plans, Undersecretary of State for Arms Control and International Security John Bolton noted in early 2004 that the Khan nuclear network went beyond Iran, Libya, and North Korea, to "several other" customers, which has been understood by diplomats to be a reference to Syria. However, the intelligence community is divided on the issue of whether Syria actually acquired nuclear technology.[38] As far as indigenous capabilities, there are no indications that Syria has the necessary infrastructure to pursue the development of nuclear weapons.

Traditionally, Syria has kept a very low profile on its WMD capabilities, and has made only sporadic references to this issue over the years. When Syria's ambassador in Cairo was quoted in late 1996 as saying that Syria would respond to a nuclear attack from Israel with chemical weapons, a prompt denial by the ambassador was issued the following day, claiming that he had in fact not made any reference to chemical weapons or to any threat to use them.[39] However, in an interview with the *Daily Telegraph* in early 2004, President Assad did say that because of Israel's belligerent behavior, "it is natural for us [Syria] to look for means to defend ourselves. It is not difficult to get most of these weapons [WMD] anywhere in the world and they can be obtained at any time."[40] He added that in order for Syria to agree to destroy its capabilities, Israel must abandon its nuclear capabilities first.

The war in Iraq put Syria in a difficult predicament. Syria chose to oppose the war and thereby placed itself in opposition to the US. Following the war, the US accused

Syria of hiding Iraqi WMD, and since the end of the war, these allegations have featured repeatedly in references made by the Bush administration to Syria. No hard evidence has been produced to support this claim, although David Kay, in an interview upon resigning as head of the Iraq Survey Group, was quoted as saying that he had uncovered proof that part of Saddam Hussein's secret weapons program was hidden in Syria shortly before the war.[41] In any case, in light of increased attention on the proliferation of WMD in the Middle East, there has been growing international pressure on Syria to take clear steps with regard to its suspected WMD capabilities. The US and the Europeans have increased their efforts to alter Syria's behavior, explaining that, as in the case of Libya, this was the road to better relations with the US.

Despite friction between the US and Syria, the US did not place Syria in the group of states comprising the axis of evil[42] and waited for Syria to "get the message" regarding US expectations of cooperation.[43] That is to say, the US anticipation was for Syria to modify its support of terrorism, its presence in Lebanon, and its WMD programs and thus comply with US demands. Syria earned some positive reactions from the administration when it seemed that Assad was responding to US pressure with his steps to close the border with Iraq in order to deny entry to extremists who hoped to fight in Iraq against the Americans, and with his call for the renewal of negotiations with Israel.

However, these steps were not sufficient to fulfill US demands. Therefore, the US increased its pressure on Syria, culminating in the Syria Accountability and Lebanese Sovereignty Restoration Act of 2003.[44] According to this act, diplomatic and economic sanctions on Syria would remain in effect until Syria abandoned its support of terrorism, withdrew its forces from Lebanon, and stopped its WMD programs. Bush was given the discretion to choose which sanctions to impose, and he announced these by executive order in May 2004.[45]

Europe has also demonstrated increasing concern with Syria's actions, with Britain coordinating its efforts with the US. The US and Britain presented a united front toward Syria, stating that it must take steps to abandon its WMD programs even if at present Israel retained its nuclear capabilities. They noted their support for a weapons of mass destruction-free zone (WMDFZ) in the Middle East, but that some states, such as Syria, had to disarm first.[46] This position acknowledged Israel's unique situation and rejected Syrian attempts to link dealing with its own WMD programs to what it expects should be done with respect to Israel. It was clarified to Syria that this stance would not be accepted as an excuse for avoiding taking immediate concrete steps.

In this context, Britain stated that it planned to lead a European effort aimed at pressuring Syria to abandon its programs. Britain was attempting to follow up on its successful involvement in the Libyan case, but this time with the cooperation of France and Germany, similar to the October 2003 diplomatic efforts toward Iran.[47] In addition, the EU has conditioned signing a significant bilateral trade agreement with Syria on indications of a clear Syrian commitment to uphold international arms control agreements.[48] Britain, Germany, and the Netherlands took the lead in opposing the EU signing an Association Agreement with Syria as part of the Barcelona Process before Syria agreed to sign and adhere to WMD non-proliferation treaties. The agreement with Syria was initialed in December 2003, but it included a WMD conditionality clause.

Syria became the third Middle Eastern country that engaged the Europeans on proliferation. The issue's growing prominence on the European agenda after the cases of Iran and Libya planted it squarely as a focus of European political efforts.[49] For its part, Europe emphasized diplomacy as the primary means to convince these states to abandon their WMD programs.

➤ Saudi Arabia

Saudi Arabia does not have a nuclear weapons capability, and there is no evidence to suggest that it has the necessary technical infrastructure to develop one.[50]

However, there are some indications that Saudi Arabia might be considering purchasing a capability or acquiring a nuclear guarantee. Two developments of recent years have marked Saudi Arabia as potentially more inclined to become a proliferator than before: Iran's persistence in its efforts to achieve a nuclear weapons capability, and fears that the US might remove its security assurances from Riyadh due to increasing US criticism of the Saudi monarchy, notably since 9/11. These developments have nurtured a growing drive on the part of Saudi Arabia to win new security reassurances. An October 2003 editorial in the Saudi *A-Sharq al-Awsat* argued that Saudi Arabia feared the Iranian bomb, which constituted a threat to Saudi Arabia more than to Israel.[51]

These concerns and the shifting sands in the Middle East may have prompted the Saudis to review their nuclear position. According to the British *Guardian*, Riyadh was considering three alternatives: acquiring independent nuclear capabilities; maintaining or entering into an alliance with a nuclear power; or promoting a nuclear-free Middle East.[52] Other reports have mentioned Saudi efforts to reach a nuclear agreement with Pakistan: a deal whereby Pakistani nuclear warheads would be placed on Saudi soil.[53] The uncovering of the Pakistani nuclear network has increased speculations regarding Saudi-Pakistani cooperation in the nuclear realm. Additional information has suggested that Saudi Arabia helped finance the Pakistani nuclear project,[54] which implies that Saudi Arabia may have been planning for this contingency; i.e., that it financed Pakistani capabilities as a future guarantee of Pakistani protection, in the context of a strategic alliance. Further evidence of possible Saudi advance preparations is the purchase of intermediate range ballistic missiles back in the 1980s, which seemed intended primarily for non-conventional scenarios.[55]

Overall, while these directions are as yet highly speculative, Saudi Arabia has signaled a degree of insecurity, intensified by the recent American criticism leveled at it. This suggests that the US might deal with this state by means of a two-pronged approach that would reassure Riyadh and provide it with sufficient security guarantees on the one hand, and would set clear expectations regarding Saudi behavior in the fields of terrorism and WMD proliferation on the other hand.

➤ Egypt

Information on possible Egyptian involvement in Libya's WMD programs surfaced as part of the Libyan investigation, raising speculations as to possible Egyptian plans for acquiring nuclear weapons. According to recent reports, the US and British inspectors in Libya found evidence that might suggest that Libya "was both the source for and recipient of nuclear and missile technology and expertise from Egypt."[56] Still uncertain, however, is how incriminating the evidence found in Libya actually was. Significantly, these initial reports have not elicited notable reactions in the US or Europe. Indeed, verified information of this sort would signal an Egyptian departure from its stated position on the nuclear issue. It is commonly assumed that Egypt made a strategic decision not to pursue a nuclear weapons option when it ratified the NPT in 1981. Egypt has also invested much energy over the years in promoting the idea of a WMDFZ in the Middle East.[57]

However, the debate within Egypt over the nuclear issue never died, and the idea of Egypt becoming a nuclear state remained rooted in parts of the elite. Some analysts maintain that Egypt has the necessary means and know-how to develop a nuclear project, and what is lacking is the political will to do so. Mubarak was quoted in 1998 as saying that if the time came and Egypt needed nuclear weapons, it would not hesitate to develop them. Such references convey the sense that, should it so desire, Egypt has the means be a viable player in the nuclear arena.[58]

While the implications of the newest reports are still murky, it can be speculated that the option of buying off-the-shelf nuclear capabilities could have made sense for Egypt, especially if it could promote a nuclear project on Libyan soil without violating the NPT or clashing with the US. Such an option of "a bomb in the basement of your neighbor," assuming the program had proceeded as planned, could have provided Egypt with a nuclear option when deemed necessary, without exacting a diplomatic price. In light of Libya's disclosures, even if there was such an option, this route was eliminated.

➤ Israel

Ambiguity in the nuclear realm has been Israel's highly guarded policy since the 1960s. However, non-Israeli sources over the years have cited estimates of Israel's assumed nuclear capability – both numbers of assumed warheads, as well as particulars relating to the history of the development of Israel's nuclear program and its nuclear infrastructure.[59]

There were no developments of note in the 2002–2004 period regarding Israel's own WMD capabilities. However, the developments in other Middle East states sparked some renewed demands, notably from Syria, for equal treatment of Israel in the WMD realm. So far, such demands for Israel to deal with its nuclear option or to advance in the direction of the creation of a nuclear weapons-free zone in the Middle East have not led to concrete action. In fact, as noted, both the US and the EU have refused to accept the basis for Syria's conditional approach, namely, that "when Israel deals with its capabilities we will deal with ours."

Assuming that the current trend in confronting WMD threats in the Middle East continues, pressure could increase for some steps to be taken with regard to Israel as well. This could take the form of pressure to join the NPT, perhaps with an innovative mechanism of inclusion that would likewise allow India and Pakistan to join,[60] or initiating regional dialogue on arms control and the creation of a WMDFZ in the Middle

East. The latter option would encourage a focus not only on weapons, but also on regional security concerns. If current efforts to contain Iran fail, such a dialogue framework could become all the more urgent.

➤ REFLECTIONS ON PROLIFERATION

The analysis above highlights three themes relating to WMD in the Middle East: the centrality of Iran to an evaluation of the overall unfolding proliferation trends; the emerging international positions in dealing with WMD threats; and the long-term effort not only to strengthen international arms control efforts but to create a regional security architecture that would address motivation and security concerns in a serious and comprehensive manner.

That Iran assumed a prominent place in this analysis reflects the complexity of the case. Opinion is divided over whether Iran has in fact been cooperating with the international community or not, and how meaningful this cooperation has been. Clearly a full answer to this question necessitates a much closer examination of the internal situation in Iran, which is beyond the scope of this chapter. Nevertheless, the weight of evidence as presented here does not suggest a change of direction in Iran, but rather a sense that Iran is playing for time, especially when compared to the cooperative stance adopted by Libya. In this sense, Iran's cooperation has been indicative of a desire to ward off harsh measures in the short term, while inching toward the goal of nuclear weapons capability.

Beyond the confines of Iran itself, however, the interpretation of Iran's stance is pivotal for an overall assessment of the current role of WMD in the Middle East. If Iran is placed in the "cooperative corner" together with Libya, then it seems as if WMD might be receding in importance as a threat to the region, and that Syria could be the next stop on the road to a WMD-free Middle East, now potentially an achievable goal. If however one is more skeptical of developments in Iran, the case of Libya then tends to stand out more as the exception to the rule than the model to be imitated, and proliferation trends seem much more of an ongoing concern for the region. While the evidence presented here points to the latter position, the overall trend in the Middle East regarding any real change in the perceived value of WMD for regional states remains to be clarified.

As to confronting WMD threats, the international community will have to come to terms with the implications of the role that strong international actors have assumed. There has been evidence of heightened motivation of the US and European states to act, and to act in a coordinated manner. Arms control is assuming new forms, and the role of military force (whether executed or threatened) in this endeavor, on the one hand, and new forms of multilateral diplomacy on the other hand, will have to be closely followed.

Indeed, the role that has been assumed by strong states is likely to become more and more a part of the "arms control environment." International treaties were always dependent on strong global powers to assess the evidence and to decide what should be done. But as the case of US and British inspections in Libya demonstrates most pointedly, these states may likely assume some of the tasks that were previously carried out exclusively by international agencies (like the IAEA). How this will ultimately affect global arms control trends is yet to be seen, but these are significant developments that need to be monitored. At the same time, a parallel effort is needed to strengthen the global regime itself.

The very fact that the US and Britain have assumed a stronger role is coupled with the differential approach they have adopted, namely, applying the logic that different states should be dealt with through different strategies. This approach has been played out not only with regard to the choice between military force and diplomacy, but in the different modes of advancing the diplomatic route. Of note is the difference between the three-sided secret and focused negotiations that were held in the case of Libya, which kept the IAEA out of the picture, and the diplomatic track pursued in Iran, conducted very much within the framework of the NPT and IAEA. A harsher line than before has been adopted with regard to Syria, especially by Europe. Finally, the answer that Assad received from the US and Britain when attempting to link his efforts with those of Israel was a clear indication of the intention to deal with different cases differently. This approach diverges from the principle of equality that is embedded in the global non-proliferation regime and that provides the normative basis for Assad to draw this linkage between treatment of Syria and Israel.

Finally, beyond strengthening the global arms control regime, the reasons why states seek to proliferate and view WMD as a necessary component of their security must be addressed. Even if the important efforts underway to reverse proliferation trends are highly successful, there may still be some states in the Middle East that prove resistant to these efforts. As noted in reference to Israel, for these cases other means for creating a more stable regional environment might be pursued such as initiating regional security dialogue among states in the Middle East, as was attempted in the early 1990s. Such dialogue could improve the regional environment by creating long-term mechanisms for dealing with states' threat perceptions and security concerns – which lie at the basis of their attempts to develop or acquire weapons of mass destruction.

Notes

1   White House Press Release, Office of the Press Secretary, "Principles for the Proliferation Security Initiative," Washington, D.C., September 4, 2003.
2   Emily B. Landau, "The NPT and Nuclear Proliferation: Matching Expectations to Current Realities," *Strategic Assessment* 6, no. 4 (2004): 32–36.
3   Sharon Squassoni, "Closing Pandora's Box: Pakistan's Role in Nuclear Proliferation," *Arms Control Today*, April 2004.
4   Peter Slevin, John Lancaster, and Kamran Khan, "At Least 7 Nations Tied To Pakistani Nuclear Ring," *Washington Post*, February 8, 2004.
5   Tenet is quoted in *The Guardian* as saying: "Working with our British colleagues, we pieced together the picture of the network, revealing its subsidiaries, its scientists, its front companies, its agents, its finances, and manufacturing plants on three continents. Our spies penetrated the network through a series of daring operations over several years." Ian Traynor, James Astill, and Ewen MacAskill, "'Supermarket' Trade in Nuclear Technology Alarms UN Inspector," *The Guardian,* February 6, 2004.
6   "UN Calls Nuclear Sales the 'Tip of an Iceberg,'" *International Herald Tribune*, February 6, 2004.

7 John J. Lumpkin, "Iran's Nuclear Weapons Program Growing at Secret Sites, Rebel Group Alleges," *Associated Press*, August 14, 2002.

8 President Khatami told a World Council of Churches meeting in Geneva on December 11, 2003 that nuclear weapons are un-Islamic and that "we [the Iranians] cannot have nuclear weapons because in Islamic wars, in the rules of warfare, there are so many recommendations for the fair treatment of the enemy." Iran, however, has a right to develop nuclear power plants to fuel its economy, he added. See "Iranian President Reaffirms that his Country Cannot Possess Nuclear Weapons," *BBC Worldwide Monitoring*, December 12, 2003; and "Iran's President Urges More Effort to Tackle Religious Hatred," *Associated Press*, December 11, 2003. For a comprehensive review of the Iranian nuclear project see Ephraim Kam, *From Terror to Nuclear Bombs: The Significance of the Iranian Threat* (Israel: Ministry of Defense Publishing House and Jaffee Center for Strategic Studies, 2004).

9 *Middle East Newsline*, June 20, 2004.

10 IAEA Director General, "Implementation of the NPT Safeguards Agreement in the Islamic Republic of Iran," June 6, 2003 (GOV/2003/40).

11 IAEA Board of Governors, "Implementation of the NPT Safeguards Agreement in the Islamic Republic of Iran," September 12, 2003 (GOV/2003/69).

12 See "Implementation of the NPT Safeguards Agreement in Iran," IAEA Resolution (GOV/2003/81), November 26, 2003.

13 Nonetheless, Iran did provide its first declaration of nuclear activities in accordance with Additional Protocol obligations in May 2004.

14 Joby Warrick, "Nuclear Program in Iran tied to Pakistan; Complex Network Acquired Technology and Blueprints," *Washington Post*, December 21, 2003.

15 Deputy Foreign Minister of Iran Mohsin Aminzadeh was quoted saying that during his visit to Pakistan, "No Foreign Scientist Helped Iran in Its Nuclear Program," Paknews.com, January 15, 2004.

16 David Albright and Corey Hiderstein, "The Centrifuge Connection," *Bulletin of the Atomic Scientists*, March/April 2004, pp. 61–63.

17 Director General IAEA, "Implementation of the NPT Safeguards Agreement in the Islamic Republic of Iran," February 2004, (GOV/2004/11).

18 Craig S. Smith, "Iran Pospones a Visit by UN Nuclear Inspectors until April," *New York Times*, March 13, 2004.

19 "Iran Relents, Allowing UN Nuclear Inspectors," *Associated Press*, March 16, 2004.

20 "Iran Renews Uranium Enrichment," *Middle East Newsline* 6, no. 125, March 30, 2004.

21 Richard Bernstein, "Europeans Criticize Iran's Plan to Start Up Enrichment Plant," *New York Times*, March 31, 2004.

22 "Iran Defuses Nuke Scheme," *Associated Press*, April 7, 2004; Francois Murphy and Parinoosh Arami, "Iran Vows to Speed Up Nuclear Cooperation with UN," *Reuters*, April 6, 2004.

23 Albright and Hiderstein, "The Centrifuge Connection," p. 66 (based on calculation of 15–20 kg per device).

24 "Exiles say Iran has 2005 Nuclear Deadline," *Washington Post* (online edition), March 14, 2004. On suspicions of transfer of uranium enrichment activities to smaller sites, see Richard Bernstein, "Europeans Criticize Iran's Plan to Start Up Enrichment Plant," *New York Times,* April 1, 2004.

25 Herb Keinon, "ElBaradei Criticizes Iranian Recalcitrance," *Jerusalem Post*, June 15, 2004.

26 Robin Wright, "US Warms to Prospect of New Talks with Iran," *Washington Post*, December 30, 2003.

27 Ram Erez, "In Search of a Role: New European Efforts to Counter Nuclear Proliferation," *Strategic Assessment* 7, no. 1 (2004): 15–21.

28 For example, in the June 2003 CIA report on WMD proliferation, the main reference to Libya's nuclear efforts deals with civil use technology that may be used for military purposes. See CIA, "Unclassified Report to Congress on the Acquisition of Technology Relating to Weapons of Mass Destruction and Advanced Conventional Munitions, 1 January Through 30 June 2003."

29 Peter Slevin and Glenn Frankel, "Libya's Qaddafi Promises to Give Up Banned Weapons; Bush, Blair Hail Results of Nine Months of Secret Negotiations," *Washington Post*, December 20, 2003.

30 Patrick E. Tylor, "Libya Arms Talks Lasted Months; Qaddafi's First Move made before the Iraq War," *International Herald Tribune*, December 22, 2003.

31 An exceptional statement in the nuclear realm was made by Ariel Sharon in September 2002, when he was quoted saying that Libya may reach a nuclear device before Iran does (interview in *Maariv,* September 6, 2002). However, no further public reference was made to this issue until Qaddafi's declaration. Amos Harel, "IDF Hints that Sharon's Slip of Tongue Encouraged US Silence on Libya Contacts," *Ha'aretz*, April 1, 2004.

32 David Albright and Corey Hinderstein, "Libya's Gas Centrifuge Procurement: Much Remains Undiscovered," *The Institute for Science and International Security*, March 1, 2004.

33 Raymond Bonner, "Pakistani Said to Have Given Libya Uranium," *New York Times*, February 21, 2004, p. A1.

34 Stephen Fidler, "Libya Formally Declares Extent of Chemical Weapons Programme," *Financial Times*, March 6, 2004, p. 5.

35 Two weeks after the shipment was captured, Libya agreed to let American and British experts visit its WMD facilities. See Robin Wright, "Ship Incident May Have Swayed Libya," *Washington Post*, January 1, 2004, p. A18.

36 Peter Slevin and Joby Warrich, "US Will Work with UN Agency in Libya; Agreement on Dismantling Nuclear Program Ends Sniping over Duties," *Washington Post*, January 20, 2004, p. A13; Ian Trayor, "Rows Brews over UN Role in Libya," *The Guardian*, January 19, 2004.

37 According to a CIA report, "Damascus already held a stockpile of the nerve agent sarin [before 2003] but apparently tried to develop more toxic and persistent nerve agents . . . It is highly probable that Syria also continued to develop an offensive [biological weapons] capability." CIA, "Unclassified Report to Congress on the Acquisition of Technology Relating to Weapons of Mass Destruction and Advanced Conventional Munitions, 1 January through 30 June 2003."

38 Louis Charbonneau, "Some in US think Syria has Atomic Centrifuges," *Reuters*, May 5, 2004. In February 2004, the *Washington Post* reported that according to a Pakistani source, the US presented the Pakistanis with evidence in October 2003 that A. Q. Kahn traveled to Lebanon in the mid-1990s where he met a top Syrian official, and tried to sell him nuclear technology, Slevin et al, "At Least 7 Nations Tied To Pakistani Nuclear Ring."

39 For the original report see Cairo, *Al-Ahram* in Arabic, November 27, 1996, in *FBIS-NES-96–233*, 27.11.96. For Darwish's denial of the statement attributed to him by *Al-Ahram*, see Damascus, SANA in Arabic, November 28, 1996, in *FBIS-NES-96–231*, 28.11.96. Darwish noted that "The Arabs do not possess or threaten anybody with weapons of mass destruction."

40 Benedict Brogan, "We won't Scrap WMD Stockpile unless Israel does, says Assad," *Daily Telegraph*, January 6, 2004, p. 1.

41 Con Coughlin, "Saddam's WMD Hidden in Syria, says Iraqi Survey Chief," *Daily Telegraph,* January 25, 2004.

42 Despite the wishes of some figures in the Bush administration. See for example John Bolton's statement that Syria should become part of this axis: Michel Evans and Richard

Beeston, "US Extends Axis of Evil to Syria, Libya and Cuba," *The Times*, October 10, 2003, p. 23.

43   Walter Pincus, "For Some, Syria Looms as Next Goal," *Washington Post*, April 8, 2003.

44   Janine Zacharia, "Bush Signs Syria Accountability Act," *Jerusalem Post*, December 14, 2003.

45   Among the sanctions imposed are bans on the export of military and dual-use items to Syria and on the export of US products other than food and medicine. In his executive order, Bush characterized Syria's actions (i.e. support of terrorism, occupation of Lebanon, pursuit of WMD, and undermining of US efforts in Iraq) as constituting "an unusual and extraordinary threat" to the national security, foreign policy, and economy of the US. See: The White House, Office of the Press Secretary, Executive Order, May 11, 2004.

46   Anton de la Guardia, "UK: Syria, not Israel, Must Give Up WMDs," *Jerusalem Post*, January 8, 2004.

47   Douglas Davis, "Britain to Push Syria on WMD," *Jerusalem Post*, December 25, 2003.

48   Judy Dempsey, "Demands on WMD Threaten Syria-EU Trade Deal," *Financial Times*, February 12, 2004.

49   *Middle East Newsline*, February 25, 2004; see also *Middle East Newsline*, April 2, 2004.

50   See Wyn Q. Bowen and Joanna Kidd, *The Nuclear Capabilities and Ambitions of Iran's Neighbours*, prepared for the Nonproliferation Policy Education Center, February 2004.

51   Reported in Ze'ev Schiff, "A Nuclear Dialogue," *Ha'aretz*, October 29, 2003.

52   Ewen MacAskill and Ian Traynor, "Saudis Consider Nuclear Bomb," *The Guardian*, September 18, 2003, p. 1.

53   "MI Head: Saudi Arabia Seeking Nuclear Missiles," *Ha'aretz*, October 22, 2003.

54   David R. Sands, "Israeli General Says Saudis Seek to Buy Pakistan Nukes," *Washington Times*, October 23, 2003. See also Bowen and Kidd, *The Nuclear Capabilities and Ambitions of Iran's Neighbours*.

55   See Richard L. Russell, *Peering Over the Horizon: Arab Threat Perception and Security Responses to a Nuclear-Ready Iran*, prepared for the Nonproliferation Policy Education Center, March 15, 2004.

56   "Libyan Inspections point to Nuke Ties with Egypt," *Middle East Newsline* 6, no. 123, March 29, 2004.

57   For a review of the Egyptian position on the nuclear issue see: Yotam Feldner, "Egypt Rethinks Its Nuclear Program – Part I: Scientific and Technological Capability Vs. International Commitments," *The Middle East Media Research Institute (MEMRI)*, Inquiry and Analysis Series - No. 118, January 17, 2003.

58   For a discussion of Egypt's considerations regarding a nuclear option see Emily Landau, "Egypt's Nuclear Dilemma," *Strategic Assessment* 5, no. 3 (2002): 22–26. See also Wisconsin Project, "Egypt's Budding Nuclear Program," *The Risk Report*, 2:5, Sept–Oct 1996; and Wisconsin Project, "Egypt Nuclear, Chemical and Missile Milestones," *The Risk Report*, 6:5, Sept–Oct 2000.

59   See sources cited in Shai Feldman, *Nuclear Weapons and Arms Control in the Middle East* (Cambridge, MA: MIT Press, 1997), and Avner Cohen, *Israel and the Bomb* (NY: Columbia University Press, 1998).

60   Avner Cohen and Thomas Graham advance the idea of creating a separate protocol to the NPT that would allow Israel, India, and Pakistan to join as associate members. This could allow them to retain their programs, while inhibiting further development. ("An NPT for Non-Members," *Bulletin of the Atomic Scientists*, 60:3, May–June 2004, 40–44). See also Ephraim Asculai, *Rethinking the Nuclear Non-Proliferation Regime*, JCSS Memorandum 70 (Tel Aviv: Jaffee Center for Strategic Studies, Tel Aviv University, June 2004).

CHAPTER ▶▶▶

5

Challenges Facing Middle East Economies

PAUL RIVLIN

▶ PRINCIPAL ECONOMIC TRENDS

The most important development in Middle East economics in recent years has been the surge in oil revenues: between 1998 and 2003, they more than doubled in current dollars. As a result, Arab countries were able to increase their defense spending by over 20 percent, and Iranian spending more than doubled. Oil revenues in 2004 are likely to be higher than in 2003. Complementing the growth in oil revenues is the continued growth of the population and of the labor force. The population of the Arab world is increasing by some three million people a year and the labor force by about two million.

In the last two years, four countries experienced significant changes in economic direction. In the second half of 2003, Israel's economy began to show signs of recovery and these were reinforced early in 2004. Turkey experienced major economic improvements in 2003 and 2004 with faster economic growth and a large reduction in inflation. In 2003, Libya's economy began to recover from years of recession and was expected to benefit greatly from its shift in policy on weapons of mass destruction as well as from economic policy reforms. Egypt's economy moved in the opposite direction: economic growth slowed and accelerating inflation cut living standards and led to the reintroduction of government allocations of basic goods.

Since the spring of 2003, Iraq has continued to suffer the consequences of another war. Although the US invasion resulted in far less destruction than the 1991 Gulf War, conflict with US forces during and since the 2003 military campaign resulted in severe disruption of the economy and damage to the infrastructure. The other economy that continued to experience damage and income losses was that of the Palestinian Authority. Close to four years after the intifada erupted, the cumulative effect of the violent conflict has returned the economy to its level of nearly twenty years ago.

Paul Rivlin

➤ Oil Revenues

Between 1998 and 2003, Middle East oil income rose by $117 billion as a result of higher prices. The oil export revenues of Middle East OPEC members – Algeria, Iran, Iraq, Libya, Kuwait, Qatar, Saudi Arabia, and the UAE – rose from a low of $88 billion in 1986 to an estimated $169 billion in 2003. This substantial increase provided the oil producers with more room for maneuver and enabled them to increase government spending. Saudi Arabia, Oman, Qatar, and Iran used some of those funds to increase their military budgets. The oil producing states experienced lower budget deficits and improved their balance of payments, and their foreign reserves increased. They were also better able to fund investment in their oil sectors. Prior to 1999, the lack of local funds meant that foreign investment was thought to be the only way to develop the oil sector. The increase in oil revenues freed the oil producing states from the anxiety that they might become reliant on foreigners for financing as well as for technology.

In real terms, oil revenues in 2003 were only about 35 percent of their 1980 peak. In per capita terms, allowing for the huge increase in population, income levels were far lower than in 1980 and in some countries they hardly changed between the low of 1986 and 2003 (table 5.1).

Table 5.1   OPEC Middle East Countries, Oil Income per capita (in dollars, exchange rates of 2000)

	1972	1980	1986	2003
Algeria	350	1473	320	498
Iran	562	712	184	324
Iraq	560	4,322	658	417
Kuwait	13,690	29,270	5,556	6,857
Libya	5,674	34,744	2,102	2,086
Qatar	13,847	49,565	5,897	13,167
Saudi Arabia	2,938	23,820	2,335	2,839
UAE	12,647	39,900	7,292	7,714

Between 1980 and 2000, the Middle East states had a total income of about $2.55 trillion from oil exports, equal to some 30 percent of their GDP. This income, the largest rental income ever received by a group of states, went directly to governments and was thereupon channeled to different purposes. As a result, the economy was totally reliant on the state.

Fewer than one million people, or less than one percent of the labor force, were employed in the oil, gas, and petrochemical industries. The oil producers failed to create other sectors that would provide incomes during periods of low oil revenues. As a result, changes in national income and oil revenues were closely correlated. The failure to diversify on a sufficient scale also meant that the growth of productive employment was limited. This in turn has given rise to increasing unemployment.

➤ Demographics

Population growth rates in the region have declined, although they remain high by international standards (table 5.2). While the average number of children born to each

78

woman has declined throughout the region (with considerable variations between countries), the number of women of childbearing age has risen. Absolute increases in the population remain huge: over one million people a year in Turkey, Iran, and Egypt, each. Between 1970 and 1980, the Arab population increased by 38 million. Between 1990 and 2000, despite the slowdown in the annual rate of growth, the absolute increase was 73 million people.

Table 5.2   Population of the Middle East, 1970–2003 (in millions)

	Arab States	Iran	Turkey	Israel
1970	119	29	35	3.0
1980	157	40	44	4.0
1990	208	56	56	4.7
2000	281	64	65	6.0
2003 *	290	69	71	6.4

* UN estimate

Due to the age distribution of the population and the increase of women in the workplace, the labor force has grown rapidly and is expected to continue to do so for at least another twenty years. This presents the countries of the region with both an opportunity and a threat. The opportunity, known as the demographic gift, is that the working age population will remain relatively large while the number of dependents – children and elderly people – will remain relatively small. The dependency ratio will therefore remain low: if people find work, they will have fewer dependents to support. However, should there not be enough opportunities in the workforce, large numbers will be unemployed.

➤ Unemployment

In 2003, unemployment in the Arab world overall was estimated at between 15–20 percent. In Egypt alone, for example, according to official estimates, unemployment rose between 1988 and 1998 from 5.4 percent to 7.9 percent, or by 830,000. This was two and a half times the growth rate of either the labor force or the working age population. Among urban males between the ages of fifteen and nineteen, a politically sensitive group, unemployment increased from 173,000 (15 percent) in 1988 to 335,000 (22 percent) in 1998. In absolute terms, the number almost doubled and this took place during a period of relatively rapid economic growth. Moreover, the official definition of unemployment was very restrictive, which means that the true level was higher. The Egyptian government was unwilling to reduce employment in the public sector even if the implementation of economic reforms would ultimately lead to the creation of other jobs elsewhere. Similar considerations applied throughout the region.

Reductions in food and other subsidies resulted in riots in the 1970s and 1980s, but they were contained and political stability was not undermined. This was possible in part because the Arab states maintained massive public sector employment, both civilian and non-civilian. This in turn helped to retain the loyalty of those employed. In addition, the regimes maintained large police, military, paramilitary, and other security forces that were used against their own populations when necessary.

➤ Economic Growth

In recent years, most of the economies of the region have been growing at significantly slower rates than those required to reduce unemployment, taking into consideration the growth of the working age population and the greater percentage of this group among the population at large. Unless a sufficient number of new jobs are created, unemployment will increase beyond existing high levels, further threatening social and political stability as well as economic well being.

Table 5.3 shows that in four out of the nine Arab countries listed, national income per capita was lower in 2002 than in 1980. In Syria the increase was very modest, and only four countries experienced positive growth, the most important of which was Egypt. Iran's recovery in recent years was due to economic reforms and increases in oil revenues. Between 1980 and 2002, the best achievements were in Turkey, with an increase in national income per capita of just over 50 percent despite repeated economic crises in the 1990s, and Israel, which recorded a 38 percent rise.

Table 5.3 GDP per capita, 1980–1999 (in dollars, 1995 prices and exchange rates)

	1980	1990	1999	2002
Algeria	1,692	1,638	1,573	1,660
Bahrain	12,022	8,551	9,329	9,640 ↓
Egypt	731	971	1,194	1,268 ↗
Jordan	1,715	1,436	1,468	1,551
Morocco	1,114	1,310	1,268	1,339
Oman	3,509	5,581	5,900	6,405 ↑
Saudi Arabia	11,554	7,101	6,455	6,331 ↓
Syria	1,071	956	1,242	1,196
Tunisia	1,641	1,823	2,394	2,584 ↙
Iran	1,396	1,469	1,539	1,789
Turkey	1,046	1,377	1,544	1,581
Israel	11,987	14,217	16,734	16,546 ↗

➤ Economic Liberalization

Economic liberalization has slowed in Egypt and in the oil rich Gulf States since 2000. Fear of unemployment has increased generally, but in the oil rich states the rise in oil revenues has made change less urgent. As Arab states began the process of economic liberalization, they were constrained by the state itself, that is, the central government and all the bodies that it directly controlled. A thriving private sector, especially one linked to the international economy through the mechanisms of globalization, would reduce the power, status, and relative importance of the state. Rulers would have to share power to a greater extent with the domestic and foreign business community. When proposals for economic liberalization were put forward, therefore, a major issue was whether the state would lose power to the private sector. Insofar as a shift in the distribution of power was the likely outcome, it reduced government willingness to implement reforms. These factors did not apply in Israel and Turkey, while in Iran there were ideological as well as political pressures to limit reform.

Government in the Middle East is big and there have been no signs of it shrinking. In the late 1990s the total size of the bureaucracy in Egypt, Algeria, Iraq, Jordan, Morocco, Saudi Arabia, and Syria was between eight million and ten million positions, excluding those in public sector industries. The huge numbers of civil servants form a vested interest that is hard to dislodge. Bureaucracy's business is governing and this inevitably comes into conflict with economic liberalization that requires the decentralization of power.

➤ Information and Globalization

During the last twenty years, the exposure to information from all over the world increased significantly at a time when income gaps in the Middle East widened. Populations in most states were increasingly exposed to economic conditions in different parts of the world. Television, perhaps the most powerful medium of communication, allowed viewers to see that Arab states were lagging in the international economic race, or, for example, that in 1980, GDP per capita in Saudi Arabia was 4.8 times that of South Korea while in 2002 it was only 86 percent. Similar trends prevailed in the other oil rich states. The recognition of these increasing differences became powerful sources of frustration that contributed to the upsurge of fundamentalism in the Arab world, while in Iran they led many to call for closer ties with the West.

➤ Impact of Military Conflict

In the last two years, military conflict and low-intensity confrontation have wrought major destruction in Iraq. Elsewhere in the Middle East, the ongoing Israeli–Palestinian violence has resulted in massive damage to the Palestinian economy, due to destruction of the infrastructure and severe income losses. Between 2000 and 2003, Israel experienced its deepest recession ever, both as a result of the intifada and the collapse of the international hi-tech boom. Jordan and Egypt were also affected by the Israeli–Palestinian conflict, with negative effects on tourism revenues and on other economic sectors. The intifada and the unrest in Iraq, as well as the negative image that the Middle East has acquired in part because of the rise of Islamic terrorism, has reduced foreign investment in the region to one of the lowest in the world. Thus the Middle East (with the partial exception of Israel) has failed to benefit from one of the main engines pushing development in the world economy during the last two decades. Not only have few investments been attracted by the Arab world but also a lack of opportunities has resulted in an exit of funds.

➤ Defense Spending

Defense spending in the Middle East has been rising in recent years. Although the most recent estimates are for 2002, there is no evidence of a slowdown since then. The rise in defense spending in the Gulf was made possible by the increase in oil revenues, and the ongoing Israeli–Palestinian violence resulted in increased allocations by Israel for defense.

The Middle East states spend a disproportionate share of their income on the military. In 2000, Arab countries accounted for almost 6 percent of world military spending, but they had only an estimated 1.9 percent of the world GDP. In the period 1980–89, Arab states, excluding Iraq, spent an annual average of $55 billion measured in 2002 prices on defense. In 1991–2000, they spent an annual average of $48 billion, also in 2002 prices. The decline in the 1990s was due to the fall in oil revenues.

In 1999, oil prices started to increase and this led to an increase in defense spending in the Gulf (table 5.4). Extreme caution should be exercised in using these figures, given the problems of reporting, inflation, and exchange rates. Between 1999 and 2002, defense spending, as reflected in official defense budgets in Iran, Saudi Arabia, Oman, and Qatar, rose by $8.3 billion or by 37 percent.

Table 5.4 Defense Budgets, 1995–2002 ($ billion, current prices)

	1995	1999	2000	2001	2002
Algeria	1.23	1.83	1.88	3.15	2.10
Bahrain	0.27	0.32	0.32	0.33	0.34
Egypt	1.97	2.39	2.39	1.19	2.75
Jordan	0.55	0.75	0.79	0.79	0.78
Kuwait	3.70	3.28	3.21	3.29	3.30
Lebanon	0.49	0.57	0.58	0.92	0.81
Morocco	1.55	1.42	1.38	1.39	1.30
Oman	2.01	0.24	0.28	2.42	2.63
Qatar	0.33	1.47	1.43	1.63	1.74
Saudi Arabia	13.22	17.60	20.00	21.06	21.33
Sudan	1.39	0.42	0.57	0.57	0.63
Syria	0.96	1.00	1.07	1.27	1.36
Tunisia	0.35	0.36	0.32	0.32	0.38
UAE	1.95	1.64	1.64	1.64	1.64
Yemen	0.29	0.37	0.40	0.46	0.49
Total Arab states	30.26	33.66	36.26	40.43	41.58
Iran	2.55	1.64	2.71	4.07	4.90
Turkey	6.61	9.95	10.00	7.21	9.22
Israel	7.08	8.23	8.93	8.87	9.84
Total non-Arab states	16.24	19.82	21.64	20.15	23.96
Total Middle East	**46.50**	**53.48**	**57.90**	**60.5**	**65.54**

➤ ECONOMIES IN BRIEF

➤ Israel

At the end of 2003 and in early 2004, the Israeli economy began a long-awaited turnaround, with GDP growth of 3–3.5 percent forecast for 2004. The economy had been in a recession since 1996, largely as a result of very tight monetary policies that constricted economic development outside the hi-tech sector. In the autumn of 2000, almost simultaneously, the second intifada began and the collapse of share prices on the NASDAQ brought an end to the international hi-tech boom. The intifada led to a collapse of tourism revenues and further reduced activity in other sectors. In 2000–2003, these effects coalesced and resulted in negative national income growth at an average of -0.2 percent a year and national income per capita growth at average of -2.2 percent a year, rising unemployment, and large cuts in government non-military spending. The Bank of Israel estimated that the GDP in 2003 was about $7 billion lower than it would have been had the intifada not occurred. The cumulative losses between 2000 and 2003 exceeded $10 billion.

Finance Minister Binyamin Netanyahu called for a complete overhaul of the economy based on sharp cutbacks in the government share of the national income. This meant reductions in non-military government spending and taxation, both of which were implemented in 2003 and 2004. The finance minister also proposed to accelerate the privatization program. International credit rating agencies and forecasting organizations continued to give Israel favorable reviews, emphasizing its formidable technological capacity and skilled labor force. Macro-economic problems and misallocation of resources, however, meant that the economy's potential was not being realized and therefore unemployment remained high. Immigration has declined in recent years and so has the population growth rate, which fell from an average of 2.6 percent in 1996–2000 to 1.8 percent in 2003.

Between 2000 and 2004, Israel increased its defense allocation. Given the concomitant recession, when government revenues from taxation were falling, the increase intensified the squeeze on government civilian spending. Defense consumption rose from 8.6 percent of the GDP in 2000 to 10 percent in 2002 (the most recent year for which figures are available). If hidden costs are included, such as those of conscripted manpower, then the defense share of the GDP rose from 10.9 percent to 12 percent.

Other government cuts included reductions in welfare payments, some of which had grown rapidly over the last decade. Poverty levels subsequently increased and the distribution of income and wealth has become ever more unequal. The rise in unemployment and poverty sparked a public debate about the kind of society that Israel should be, but the debate was frequently obscured by terrorist attacks and security problems.

➤ Algeria

The Algerian economy began to experience a recovery after years of civil war. In 2003 it grew by 7.4 percent and in 2004 its expected growth was over 6 percent. This was largely the result of higher oil and natural gas export revenues. Between 1986 and 2003, these rose on a per capita basis by some 50 percent. Algeria has 1 percent of the world oil reserves and 2.5 percent of the world gas reserves. Between 1995 and 2000, oil and gas revenues accounted for 25 percent of the GDP, 96 percent of exports, and 60 percent of budgetary revenues. Algeria's dependence on hydrocarbons continued despite large investments in industry. The latter were far from successful: between 1990 and 1997 manufacturing output fell by 40 percent. As a result, the share of manufacturing in the GDP declined from 10.6 percent in 1980 to 7.7 percent in 2000. This was particularly problematic given that manufacturing was the sector that provided most hope for the development of badly needed employment.

Between 1985 and 1986 oil income fell by 50 percent. In response the government borrowed abroad, but soon found that it could not repay its debts. Between 1986 and 1998, GDP growth averaged 1 percent a year, much less than the population growth rate. In 1994, following the collapse of oil prices the previous year, there was a severe financial crisis. As a result, foreign debt repayments were rescheduled and radical economic reforms were introduced. These measures helped to stabilize the balance of payments, foreign debt, and budget deficit, but they also resulted in a recession and growing unemployment. By 1998, the GDP was at half its peak of 1986 and GDP per capita even lower.

Unemployment has reached over 30 percent, with a much higher rate among young people. Since 1986, poverty has risen sharply, and much of the middle class lost substantial sums of money. The rapid growth of the labor force (which averaged 3.4 percent a year in the 1990s) and labor shedding in the public sector (between 1990 and 1998, 500,000 lost their public sector jobs) has made privatization very problematic. Cutbacks in the public sector together with slow growth in the private sector resulted in high unemployment, which in turn increased the number of poverty stricken. In the 1990s, the active labor force increased by over 4 percent a year. In order to absorb the growth of the labor force and reduce unemployment, a GDP growth of 6–8 percent a year is needed. Given the sharp fall in investment from 39 percent of the GDP in 1980 to just under 24 percent in 2000, this prospect seems remote. Both the army and the trade unions have opposed further radical reforms.

Housing shortages and a severe lack of social and infrastructure services represented serious problems in urban areas. Although the population growth rate decreased sharply in the 1980s and 1990s, the movement of people from war torn and impoverished rural areas to the cities has resulted in the growth of shanty towns. These areas were especially vulnerable to floods and earthquakes that have, since 2000, resulted in the deaths of thousands and the displacement of hundreds of thousands. A severe drought affected the agricultural sector in 2000, heavy flooding affected northern Algeria in November 2001, and there was a major earthquake in May 2003. These natural events compounded the extant socio-economic difficulties.

➤ Egypt

Since 2000, the Egyptian economy has been in a recession. In a country with high unemployment and widespread poverty, this is a serious threat to social stability. Unemployment has increased and inflation has reduced the purchasing power of the population. In 2002, the estimated growth rate was 3 percent, half of that in 1999. Economic growth slowed at the end of the 1990s after five years of relatively fast development. The deceleration was due to the effects of the 1997 attack on tourists in Aswan, the Palestinian intifada, and the attacks of September 11, 2001 as well as domestic financial problems. The slower growth rate meant fewer new jobs were generated, which in turn accelerated growing rates of unemployment. Fears of the social and political effects of unemployment were one of the main reasons for the government's unwillingness to push forward with economic reforms, notably further privatization.

In 2000, in an attempt to strengthen the balance of payments that had been weakened by falling oil and tourism revenues as well as by lower worker remittances, the link of the Egyptian pound to the US dollar was canceled. By the middle of the year there was a cumulative depreciation of about 25 percent of the pound. The devaluation resulted in an acceleration of inflation and a further squeeze on living standards. In January 2001 a managed exchange rate was introduced with controlled devaluations, but in January 2003, the rate was unfrozen and the Egyptian pound fell again. The devaluation enabled the government to reduce what had been very high interest rates. During 2003, there were shortages of bread and other commodities along with price rises in bread and other basic items.

Egypt remains short of foreign currency, as its exports are chronically weak and it relies heavily on imports. In 2002, exports of goods came to $6.7 billion, but 93 percent were agricultural products, hydrocarbons, and minerals. Egypt, like many other countries in the region, has little to sell beyond a limited range of traditional goods and oil-related products. With imports of almost $13 billion, Egypt has little alternative but to export, yet the lack of a diversified range of exports means that import expenses cannot be covered easily. If and when there are balance of payments problems the domestic economy has to be squeezed. The 2003–2004 budget reflected these problems with increased allocations for subsidies, interest payments, and debt repayment. Expenditures were planned to rise much faster than revenues and as a result the budget deficit was 60 percent higher than in 2002–2003. In 2004, plans were announced to reintroduce rationed supplies of basic goods at subsidized prices for the poor. Unfortunately for economic progress, the implication is Egypt is taking the first steps back toward the kind of controlled economy that it abandoned nearly fifteen years ago.

In contrast to these inauspicious developments, the Egyptian government has been rethinking its policies toward Israel. At the initiative of the Egyptians and with US encouragement, contacts began on establishing a Qualified Industrial Zone in Egypt to produce products jointly with Israel. These will be exported to the US tax free within the framework of the Israel-US free trade agreement. Early in 2004, talks were held between the two countries on an agreement for jointly producing goods to be exported to the United States. Such developments are especially important for Egypt, since Egyptian exports to the US are currently subject to import duties and quantitative restriction.

➢ Iran

Iran's economic performance has improved in recent years as a result of economic reforms and high oil prices. In the period 2001–2003, national income rose by an annual average of 5.8 percent, compared to 3.8 percent in 1995–2000. Fast growth is forecast for 2004 as a result of high oil revenues. In March 2002, the multiple exchange rate system was unified and most restrictions on current account transactions were lifted. Non-tariff barriers on imports were cancelled and the tariff system was reformed. A new foreign investment law was introduced and monetary and tax reforms were introduced.

These reforms were part of a long-term reform process that began in response to the foreign debt crisis of 1993. Since then, Iran has overcome its major foreign debt crisis, although it still has one of the highest inflation rates in the region.

At the same time, the population growth rate has slowed markedly, a direct and indirect result of government policy. Between the early 1980s and 2002, the population growth rate declined from 3.7 percent to 1.4 percent, and the fertility rate fell from 6.8 to 2.6. This change was not accompanied, as elsewhere in the Arab world, by a rise in the female labor force participation rate. Iranian women are increasingly turning to education, especially higher education, as an alternative to work in the formal sector. At the same time, as a result of fast population growth rates of the 1980s, the labor force is growing at about 4 percent a year, which means that some 700,000 individuals join the job market every year, far more than the number of jobs created. In 2001, more than 40 percent of Iranians with a high school education or higher were unemployed. In 2002, about 160,000 of the most educated citizens emigrated.

In addition to the challenge of generating employment, Iran faces a number of other socio-economic challenges. These include a large bureaucracy and widespread corruption which is institutionalized through autonomous charitable organizations – *bonyads* – that have control of billions of dollars worth of resources and operate without transparency. In addition, the country suffers from massive air and water pollution, poverty, and accumulated earthquake damage.

Furthermore, since 1995, Iran has been subject to US sanctions because of its support for international terrorism. US companies and their foreign subsidiaries are prohibited from conducting business with Iran, including financing development of petroleum resources. The US Iran-Libya Sanctions Act (ILSA) of 1996 (renewed for five years in July 2001) imposed mandatory and discretionary sanctions on non-US companies investing more than $20 million annually in the Iranian oil and natural gas sectors. These measures have had a major effect in retarding the development of the oil and gas sectors and on the economy as a whole.

➢ Iraq

The 2003 war resulted in far less damage to the economy than the 1991 Gulf War, but the disruption of the war and the effects of terrorism since then have resulted in losses worth billions of dollars.

National income per capita in 2003 was estimated at about $500, compared with $3,500–$4,000 in 1980. Since 1980, as a result of wars, massive human and infrastructure destruction, economic mismanagement, international sanctions, and increasing foreign debt, the GDP collapsed while the population continued to grow. The Iraqi economy, which had shown greater agricultural and industrial potential than many others in the Middle East, has over the last twenty-five years become increasingly dependent on oil income and thus on prices determined on international markets.

In the beginning of 2004, some 60 percent of the population was dependent on food rations provided by government, as they had been under Saddam Hussein's regime. According to the World Bank, national income fell by 12 percent in 2001, by 4 percent in 2002, and by 31 percent in 2003. This accumulated fall of 42 percent must be understood against an increase in the population estimated at 6 percent, resulting in a fall in GDP per capita of about 45 percent.

The World Bank and UN have estimated that in the period 2004–2007, a total of $55 billion will be needed to reconstruct the economy. Some $36 billion will be needed to reconstruct fourteen non-oil economic sectors, and the US estimated that rehabilitating the oil sector and providing security will cost an additional $19 billion. In October 2003, a donors conference in Madrid resulted in pledges of aid totaling $32 billion, including $18.6 billion from the US and at least $5.5 billion from the IMF and the World Bank.

Following the 2003 war, oil production recovered from a low of 150,000 barrels a day in April to 1.8 million barrels (mb/d) a day in December. Average output in 2003 was 1.285 mb/d compared with 2.014 mb/d in 2002 and 2.3 mb/d in 2001. Oil export revenues in 2003 were estimated at $9.6 billion, 26 percent lower than in 2002 and less than half their 2000 level.

The Iraqi economy was run by the Ba'ath regime based on a very centralized program and a dominant public sector. At the end of 2003 public sector employment was estimated at 1.3–1.5 million, with 500,000 workers in 200 state-owned enterprises. In 2003, the US Coalition Provisional Authority announced major economic reforms. Prices for all goods except energy and food were freed of government controls; trade licensing was abolished, as were trade tariffs. A 5 percent tax on imports was introduced to help finance the reconstruction of the economy.

The Iraqi trade minister, however, voiced his reservations: the economy had, in his view, already been subject to too many "experiments" under Saddam Hussein. Privatization and subsidy cuts would curtail lowering unemployment and rising price trends. These would join big changes that have been implemented such as the opening of the economy, apart from oil, to foreign investment and the lowering of taxes on income to 15 percent and import duties to 5 percent. The reduction of import duties and the ending of other controls on imports have resulted in a flood of imports into Iraq. Insofar as these increase the supply of a wide range of goods and services needed by the economy, this was beneficial in the short term. In the long term, the problem is that local industry, handicapped by many factors, will find it hard to compete with imports. This is a phenomenon common to other Arab countries, such as Egypt, that have suffered much less than Iraq.

At present between 40 and 60 percent of the Iraqi labor force is unemployed or underemployed. The civilian labor force is estimated at about 7.5 million but with the disbanding of the army it increased by up to one million. Natural growth of 3 percent means that 250,000 people are added to the labor force each year. If state-owned companies are privatized and there are consequent job losses, then the need for creating more employment will be greater. If the private sector as well as the public sector is unable to compete with imports then there will be even more job losses.

➣ Jordan

In 2001–2003 Jordan experienced an improvement in its economic growth rate, but as the population growth rate was about 2.8 percent, this translated into only a modest rise in GDP per capita – less than 1 percent a year. Between 1990 and 1995, Jordan's economic growth rate averaged 7.6 percent per year. Between 1996 and 2000, it fell to an average of only 3.0 percent. As a result of inadequate economic growth and fast population growth, GDP per capita decreased from an annual average of $1,990 in 1980–1985 to $1,750 in 2001.

Jordan has developed one of the best environments for business in the Arab world. This was achieved by improving macro-economic stability, increasing access to the Jordanian market for foreign companies, and improving the legal framework. Since 2001, exports to the US have increased sharply as a result of the establishment of the so-called Qualified Industrial Zones and the signing in that year of a free trade agreement with the US. Jordanian exports to the US increased from $16 million in 1998 to $412 million in 2002. This made possible the creation of 40,000 new jobs, mainly in the garment industry, even while Jordan faced the intifada to its west and the buildup to the war in Iraq to the east. Preliminary figures for 2003 showed another strong increase in Jordan's exports to the US. In 2002, an Association Agreement was signed with the European Union.

Despite the development of exports, the economy remained dependent on external sources of aid, including financial and military assistance from the West and remittances from Jordanians working overseas. In recent years, these have contributed as much as 25 percent of national income. It also continued to face the poverty that has increased during the past twenty years. At the end of the 1980s about 3 percent of the population was classified as poor, but by 1992 the proportion rose to 14.4 percent, and the number of poor increased more than sixfold. After the 1991 Gulf War, the collapse of oil prices, the return of approximately 300,000 Jordanians from the Gulf States, and the subsequent drop in worker remittances exacerbated unemployment and increased the number of those living below the poverty line. In 2001, 11.2 percent of the population was below the poverty line: in absolute terms this was a larger number than in 1992. In 2003, unemployment was estimated at 15 percent with labor force growth at 4 percent a year. Unemployment among the young was much higher, and creating additional new jobs for young job applicants remained an economic, social, and political objective.

➢ Lebanon

By mid-2004, there was some evidence of a modest acceleration in Lebanon's growth rate during the period 2001–2003 with an annual average GDP growth of 1.4 percent, a slightly faster rate than that of the population. Private consumption has risen and investments in real estate have increased. This included the south of the country, where Hizbollah was under pressure from the local population not to disturb this development directly or indirectly.

Lebanon's economy began to recover following the signing in 1989 of the Ta'if accords that ended the civil war, and their implementation in 1991. Inflation was reduced from 15 percent in 1990 to 0.5 percent in 2000. This helped replenish foreign exchange reserves and restore macroeconomic stability. National income grew rapidly during the period of reconstruction from 1991 to 1997, but decelerated thereafter, with a recession in 2000. Between 1996 and 1998, growth averaged 3.7 percent a year; between 1998 and 2002 it averaged 1.5 percent.

Poverty and income inequality increased significantly after 1975 because of the destruction caused by the civil war. The slow growth in job creation, especially for lower income groups, restricted purchasing power. In addition, the import of labor from Syria and other low wage countries held down unskilled wage levels for Lebanese workers. That some 300,000 Syrians work in Lebanon is testimony to both the vitality of parts of the Lebanese economy and to the economic effects of Syrian occupation. In addition, the return of some Lebanese from abroad and their investments boosted prices for land, housing, and medical services, making them all but unaffordable for many who had been impoverished during the civil war. Reconstruction helped to create jobs, but unemployment remained high, at about 20 percent. The unemployment rate among first time job seekers is twice that level.

In 2001, the fiscal deficit equaled 17.6 percent of the GDP, a level that was feasible during the reconstruction efforts of the mid-1990s, but could not be sustained thereafter. Rising interest outlays on the government borrowing that financed the budget deficit have led to high fiscal deficits in recent years, with the total debt reaching 170 percent of GDP in 2001. In 2002 Lebanon negotiated $4.4 billion in economic aid, mainly from Europe and the Gulf. These funds provided support for the state budget and eased the foreign debt crisis.

The central government has remained weak: it has failed to implement major privatization programs and has been unable to cut public spending. Its power was limited by Syria's occupation, Hizbollah's control in the south of the country, and divisions within the Lebanese elite. The government has had only limited success in restoring the infrastructure that would allow the country's once vibrant private sector to thrive and restore Beirut to the position it held until 1975 as the foreign trade and business center of the Arab world.

➤ Libya

Libya's economic situation improved dramatically in 2003 as a result of major policy changes and higher international oil prices. Oil export revenues, which accounted for over 95 percent of Libya's hard currency earnings (and 60 percent of government receipts), increased from some $6 billion in 1998 to an estimated $13.4 billion in 2003.

In June 2003, Qaddafi announced major policy changes. These included privatization of the oil industry and other sectors. He stated that the public sector had failed and should therefore be abolished, and he pledged to bring Libya into the World Trade Organization (WTO). In September 2003, the UN Security Council lifted sanctions against Libya following Tripoli's acceptance of responsibility for the 1988 bombing of Pan Am flight 103 over Lockerbie, Scotland. In January 2004, the US government announced that its sanctions would remain in place, despite Libya's announcement the previous month that it would end its weapons of mass destruction programs and permit unconditional international inspections of its facilities.

The economy depended heavily on oil and remained largely state controlled. In 2002, after some thirty years of socialism, the government began to liberalize the economy. On January 1, 2002, the dual exchange rate system was unified at a fixed rate and the foreign exchange rationing and import licensing requirements were eliminated.

Although the population growth rate slowed from 4 percent a year in the 1970s and 1980s to 3 percent in the 1990s, the growth of the labor force in the 1990s exceeded 3.5 percent. Between 1997 and 2002, economic growth averaged 0.8 percent a year, far too low to enable all those joining the labor market to find jobs. Furthermore, much of the growth was in the oil sector, which employed very few people. As the private sector did not absorb more labor, the government was forced to provide jobs. In 2001, 53 percent of Libyan employment was in the public sector. In large part due to higher oil export revenues, Libya experienced strong economic growth during 2003, with the gross domestic product estimated to have increased by 5.6 percent.

Foreign investment and economic growth have been discouraged by a crumbling infrastructure and a lack of spare parts and foreign technology. These were the results of sanctions that Libya claimed cost the economy $33 billion. It has also suffered from unclear legal structures, arbitrary government decision-making, inflation, corruption, a large black market, a bloated public sector, and the huge costs of the "Great Man Made River" project.

➤ Morocco

In the period 2001–2003, economic growth averaged 5 percent a year. Agriculture plays a major role in the economy, and the rainfall levels that permitted an increase in agricultural production yielded significant positive effects on the economy as a whole. Industrial production also rose and this resulted in job creation and a fall in the urban unemployment rate from 22 percent in 1999 to 18 percent in 2002.

In November 2002, Morocco suffered heavy rains and severe flooding, which caused major infrastructure damage near Casablanca. This flooding occurred in two consecutive years, following three years of drought, during which the country's large agricultural sector (20 percent of the GDP and more than 40 percent of the workforce) had been adversely affected. Nonetheless, the economy has grown despite the three successive years of drought, mainly because of tourism receipts and remittances from Moroccans abroad. Privatization proceeds have also helped the balance of payments and the government budget, but these were largely from a one-off telecommunications sale.

One of Morocco's biggest challenges was the poverty rate. The share of the population living in poverty fell from 21 to 13 percent between 1984 and 1992, but since then it has increased, reaching 19 percent in 2000. This was mainly due to low economic growth, in turn the result of failures in agriculture and the continued growth of the population, the decelerating rate notwithstanding. Over 25 percent of the rural population was below the poverty line, compared to just 12 percent of the urban population. The severity of poverty was also much higher in rural areas, and has increased significantly since 1991.

In March 2004, the United States and Morocco reached agreement on a bilateral free trade agreement. The two sides expect the agreement to be signed and ratified in 2004. The agreement is part of President Bush's vision of creating a Middle East Free Trade Area by 2013.

➤ The Palestinian Authority

The economy in the territories nominally under the control of the Palestinian Authority has suffered massive damage since the outbreak of the violence in September 2000. This resulted from physical damage to the social and economic infrastructure, job losses in the territories and in Israel, and the loss of income from production. Damage to the infrastructure was estimated at $1.1 billion. National income fell by 5.6 percent in 2000, by 24 percent in 2001, and by 22 percent in 2002, a cumulative drop of 44 percent. Given that the population grew by up to 4 percent annually, the fall in per capita income was 50 percent. According to the UN, the Palestinian economy has foregone the growth of fifteen years and has reverted to its 1986 level. Employment fell by 17 percent between 1999 and 2002 and unemployment rose from 12 percent to 31 percent.

The budget rose as a share of national income as the latter fell. The 2002 deficit of $914 million was financed by donations from Arab, EU, and World Bank sources. The 2003 budget, also funded by aid and the release of Palestinian funds by Israel, had a deficit of $889 million. The 2004 proposed budget has a deficit of $888 million. Borrowing also financed the budget deficits and resulted in an accumulation of debt. At the end of 2002 debt came to $855 million, or between 29 percent and 37 percent of the GDP.

➤ Saudi Arabia

Saudi Arabia has been a major beneficiary of the rise in oil prices in recent years. In 2003 revenues were estimated at $81 billion, 23 percent more than in 2002. As a result, in 2003 the economic growth rate accelerated to an estimated 4.7 percent compared with an average rate of 1.2 percent in 2001–2002.

The government budget remained weak as a result of large current spending and the accumulated effects of a large internal debt. This was due to the fact that there was no income tax in Saudi Arabia and overall tax collection is very low. In order to finance the excess of spending over revenues, which derive mainly from oil exports, the government has forced banks, companies, and even wealthy individuals to buy government bonds. The interest payments on these bonds form an increasing share of government spending. The budget is forecast to move from a surplus of $12 billion in 2003 to a deficit of $8 billion in 2004 despite current high levels of oil income.

The increase in oil revenues has reduced pressure on the government to implement reforms. These are needed, as the economy has not generated enough jobs to employ the large number of young people who enter the labor market every year. In 1999 unemployment was officially estimated at 8 percent and in 2002 it was unofficially estimated at 12 percent. In 2003, a Saudi prince and prominent businessman estimated it at 30 percent. Among those entering the labor force, the rate is much higher.

Saudi Arabia has one of the fastest growing populations in the world, despite a decline in fertility and the fall in the population growth rate during the last ten years. The population, like that elsewhere in the Middle East, is very young and as a result the number of Saudis working is only 2.8 million out of a total Saudi national population of about 19 million. The number of Saudis in the private sector is 2.2 million and some 600,000 are civil servants. The foreign workforce equals three million, the vast majority of which is employed in the private sector. Attempts to reduce the number of foreigners working in the country have been half-hearted.

➤ Syria

Between 1999 and 2003, the Syrian economy grew by an average annual growth rate of 0.8 percent. Given that the population increased by 10 percent during that period, GDP per capita fell by an average of 1.7 percent a year. Economic growth slowed down in the late 1990s as the economic reform program lost momentum and the country was affected by drought. In 2000, higher oil prices and good harvests resulted in a sharp upswing, but this trend did not continue.

Oil output has continued to decline due to technological problems and depletion of oil reserves. In 1996, it peaked at 590,000 barrels per day (b/d), and in 2002 it was an estimated 525,700 b/d. Oil production was expected to continue its decline, while domestic consumption rises, leading to a reduction in oil exports. As a result, Syria is expected to become a net oil importer within a decade. In recent years, the illegal import of oil from Iraq enabled Syria to increase its own exports. Oil export revenues rose from $1.6 billion in 1998 to $3.6 billion in 2000. After the 2003 war, the US closed the oil pipeline from Iraq to Syria and so the Syrian economy will lose up to $1.6 billion annually.

Syria's economic potential in manufacturing, tourism and other services, and in agriculture has never been realized as a result of economic mismanagement. Political bureaucratic regulations have stifled economic activity, as has the lack of a proper banking system. Reforms have been partial and need to be much more extensive if economic growth is to be stimulated and jobs created on the scale needed. Unemployment is officially estimated at 10 percent and unofficially at 20 percent. Among the young, the rate is much higher.

Syria, listed as a sponsor of terrorism since 1979, has been subject to US sanctions. These sanctions prevent the sale to Syria of any goods that could have dual use and which could be used for military or terrorism purposes. US government assistance to Syria is also forbidden. Transactions with the government of Syria by US citizens and companies are banned, although there are US investments in Syria's gas sector and in other economic activities. In May 2004, US sanctions against Syria were strengthened under the Syria Accountability and Lebanese Sovereignty Restoration Act of 2003.

➤ Turkey

In 2003, the Turkish economy experienced a strong recovery from the crises of 2000 and 2001. Economic growth reached 5 percent in 2003 and in the autumn, inflation fell below an annual rate of 25 percent for the first time in over twenty years. Although growth is likely to be slower in 2004, initial trends show inflation decelerating further and the balance of payments strengthening. The fact that the country has a unified government, following a decade or more of weak and divided coalitions, has added to confidence at home and abroad. After the 2002 election, the initial policy declarations of the new government reinforced confidence: it emphasized that it would pursue Turkey's application to join the EU and would retain a pro-Western orientation. Notwithstanding sharp disagreements with the US over the 2003 war in Iraq, the economy continued to improve in 2003. This occurred despite the expected losses of transport fees for Iraqi oil and Iraqi markets for Turkish goods as a result of the war.

Between 1991 and 2000 the economy grew by an average annual rate of 3.7 percent, with sharp fluctuations, rapid inflation, and financial instability. In 1999 there was a sharp fall due to two severe earthquakes, and in February 2001 the economic situation rapidly deteriorated as a financial crisis forced a massive devaluation of the Turkish lira. Inflation and unemployment rates soared in 2001 and the GDP fell by 7.4 percent. Per capita income fell even further.

This crisis was the result of balance of payments weaknesses, problems in the banking sector, and political instability. The terrorist attacks of September 11, 2001 deepened the crisis by lowering exports and tourism revenues, reducing access to international financial markets, and weakening privatization and foreign direct investment prospects.

Turkey faces numerous economic challenges, including a large black economy and greater income inequality, both within urban areas and between urban and rural areas. The public sector is bloated and much remains to be done to modernize legal and administrative procedures. One of the main results has been a lack of foreign investment that has limited the economy's ability to generate employment. In preparation for EU membership, the government is pursuing economic as well as social reforms that are expected to encourage modernization gradually.

➤ **UNDERLYING PROBLEMS IN ARAB ECONOMIES**

The Arab economies are dominated by rents, mainly from oil. These largely accrue to governments that use them to provide employment in the public sector and the military. As a result, economic power is very centralized and resources are allocated inefficiently on a political basis rather than according to economic criteria of free markets.

The other side of government dominance is the weakness of the middle class. This has been a key feature in the economic and political development of the Arab world and has a long history. The Arab regimes that came to power in the middle of the twentieth century often made the political and economic position of middle classes even weaker. This was true in all the countries that experienced Arab socialism and nationalized assets. As entrepreneurship is associated with the middle class, the weakness of the latter has had negative effects on the economy.

Arab economies are among the least integrated into the international economy. Their main export is oil. Apart from that they export a limited range of agricultural and processed goods. The share of manufactured industrial goods in total exports is one of the lowest in the world.

The lack of trade and economic cooperation between countries in the region is a reflection of the balance of power between the state and the private sector. Governments unwilling to loosen control do not want the private sector to embarrass them by getting involved in business with controversial partners such as Israel. As a result, mutually beneficial trade and investment is prevented, not only with Israel but also between other countries in the region and elsewhere. This is one more mechanism that reduces the exposure of Arab economies to competitive forces that might over time help them develop.

Israel and Turkey have chosen to link their economies with the West. For political as well as economic reasons, the development of their relations with the EU has not been easy, but both countries have industrialized and increased their manufactured exports. The Arab states and Iran lag far behind. Much of their manufacturing sector is unable to cope with potential competition from the EU. As a result, Arab countries such as Egypt have delayed signing free trade agreements, while the status quo does not provide the incentives or mechanisms to push their economies forward.

CHAPTER ➤➤➤

6

The Egyptian Armed Forces

SHLOMO BROM AND YIFTAH S. SHAPIR

The peace treaty between Egypt and Israel, signed in 1979, is now twenty-five years old. Overall, the peace treaty has remained stable, successfully withstanding difficult tests of military confrontations between Israel and other Arab nations during this period. Criticism has occasionally been voiced in Israel concerning the nature of the Egyptian–Israeli peace, described as cold and lacking the economic, cultural, tourism, and sports relations that normally exist between friendly nations. The uneasiness prevailing in some circles in Israel regarding the durability of the peace treaty with Egypt has complemented sensitivities regarding developments in the Egyptian armed forces.

Since the peace treaty was signed and thanks to its stability, Egypt has received large allocations of money in the form of American military aid every year. This aid has enabled the Egyptian armed forces to undertake a continued military buildup in recent decades in contrast to most other Arab countries, whose military growth was arrested during this period as a result of insufficient resources and the sanctions imposed on some of these countries. As part of its buildup, the Egyptian military has gradually procured Western weapon systems that are among the state of the art systems in existence, while steadily retiring the outdated Soviet armaments still in use. In addition, the Egyptian military has been maintaining close security contacts and conducting joint military exercises with countries in the region and with Western countries, especially with the United States, which is Egypt's principal weapons supplier.

➤ THE ORDER-OF-BATTLE

➤ Ground Forces

The Egyptian armed forces, which have remained the same size in recent years, are quite large, comprising 450,000 regular soldiers, of whom 320,000 serve in the ground forces. The ground forces include two armies and twelve divisions: four armored divisions, seven mechanized divisions, and one infantry division. The army operates approximately 3,000 battle tanks. Its modern armored corps includes some 650 M1A1 Abrams tanks, which were assembled in Egypt. The rest of the armored corps is older and includes M60A1 and M60A3 US-made tanks, and 1,050 still more outdated Soviet-made tanks. This combination of old Soviet-made systems with Western, primarily American, systems – a result of the slow process of systems replacement – is also a feature of the army's other weapon systems, such as armored fighting vehicles (AFV) and artillery. Buildup has focused on the slow replacement of obsolete weapon systems by up-to-date systems at an average of about sixty tanks per year.

Over the past decade, the army obtained the advanced M1A1 Abrams battle tank, with the tanks assembled in Egypt from components delivered by the US manufacturer. The first assembly contract, which lasted from 1991 until 1998, included 555 tanks. Since 1999 the Egyptian industry assembled 100 additional tanks, and the next batch of 100 tanks is under assembly and is scheduled for completion in 2005. A procurement contract was recently signed for an additional 125 tanks, scheduled for completion by 2008. When these are finished, the Egyptian armed forces will have 880 Abrams tanks.

The slow rate of processing new tanks, however, requires the preservation and upgrading of some of the old order-of-battle, and the army was therefore interested in upgrading about 200 of its obsolete T-54, T-55, and T-62 tanks. Negotiations were underway with the OMSK design bureau (KBTM) in Russia, which has developed a new turret with a fire control system for these tanks, and which has been adapted for a Western 120–mm caliber gun. In addition, Egypt was negotiating with a company in Belarus for the upgrading of old BTR-50 armored personnel carriers (APCs), also Soviet-made. Here too only 250 APCs were involved, and the prototype of the upgraded APC has already been successfully tested in Egypt.

In artillery systems, Egypt ordered twenty-six multiple launch rocket (MLRS-ER) systems at the end of 2001, including all their accessory systems, i.e., command and logistics vehicles. The Bush administration recently approved the transfer of another 201 M109A2 and M109A3 self-propelled guns from US army drawdown. These self-propelled guns are to be upgraded before being delivered to Egypt. Once they are received, Egypt will have approximately 400 M109 self-propelled guns and another 124 SP122 self-propelled guns (a Russian D-30 gun on an M109 hull). Even then, most of the order-of-battle, about 1,000 guns, will be composed of towed guns.

➤ Air Defense

The Egyptian air defense system, an independent branch of the Egyptian armed forces, remains based mostly on obsolete Soviet systems, to which Western systems, not the top of the line, have been added. Only sixteen batteries of Improved Hawk surface-to-air missiles (SAM) have joined the main order-of-battle, which consists of ninety-three batteries of SA2 and SA3 missiles. The order-of-battle of mobile surface-to-air missiles, which is used to protect the ground forces, has undergone more substantial renewal. The fourteen SA6 batteries have been supplemented by twelve French Crotale batteries, eighteen Skyguard batteries consisting of anti-aircraft guns and Sparrow missiles, and eighty Chaparral launchers. Significantly, however, the performance of all these added mobile systems is inferior to that of the SA6 missiles they are designated to complement and replace. Another project involves upgrading the outdated SA3 missiles. Its launchers are being installed on trucks, giving them self-mobility, and its guidance systems are replaced by modern digital electronics.

➤ Air Force

The Westernization process is more prominent in the Egyptian air force than in other branches of the Egyptian armed forces. The backbone of the air force is a fleet of 217 F-16s, most of which are F-16C/D-types. It also has relatively advanced AH-64A Apache attack helicopters, as well as a number of Mirage 2000 and Phantom F-4E jets. Egypt, however, still maintains over 200 obsolete planes from the 1960s, various models of the MiG-21 and the Mirage-5.

The most prominent reinforcements in the past three years were the Peace Vector V and Peace Vector VI deals. The air force received 21 F-16C/Ds in the fifth deal, signed in 1996, and an additional 24 F-16C/D (block 40) planes in the recently completed sixth deal. In addition to procuring these new planes, Egypt was considering an upgrade of F-16A/Bs delivered in the early 1980s. Another deal involved the thirty-five AH-64A Apache helicopters that Egypt received in 1995. In this transaction, the helicopters would be converted, or "reconstructed," into AH-64Ds. However, the US administration did not approve Egypt's request to purchase the Longbow radar, which gives the helicopters superb target identification and gives the crew excellent situation awareness and acquisition capabilities, thereby constructing a battle image for the helicopter team.

Less important deals by the Egyptian air force included the upgrading of five Hawkeye aircraft to the US navy's Hawkeye 2000 standard, plus the purchase of a sixth plane of this model from US navy surplus stocks; and the upgrading of six CH-47C helicopters to the CH-47D standard. In addition, the air force has already been armed with small numbers of air-to-surface armaments with precision guidance systems, most of which are Maverick short-range missiles.

➤ Navy

The Egyptian navy is the second largest navy in the Middle East after the Turkish navy, though it is divided between two different theaters, the Mediterranean Sea and the Red Sea. The navy uses Soviet and Chinese-made ships along with American and other Western-made ships. Most of the navy's strength comes from ten frigates (eight Western and two Chinese) and nineteen fast missile boats (eleven Western, four Chinese, and four Soviet-made), plus four obsolete former Chinese submarines. The main advance in the Egyptian submarine fleet was the deal to acquire two Moray submarines. Egypt planned to buy these submarines with foreign military sales (FMS) funds. Since the US does not manufacture diesel submarines, however, a consortium was formed to produce the submarine in the US along the Dutch Moray model. Although Egypt signed a letter of intent in 2000, the contract is currently dormant.

Egypt's submarine order-of-battle will therefore continue to rely solely on obsolete Chinese submarines in the coming years.

The Egyptian surface fleet is one of the most important areas in the Egyptian military power buildup. In the course of this buildup, Western and Chinese-made weapons have almost completely replaced Soviet-made equipment. During the 1990s, the Egyptian navy obtained two Knox frigates and four Oliver Hazard Perry frigates from US navy drawdown. Another Egyptian navy deal with the US, involving the procurement of four Ambassador Mk III high-speed patrol boats, was signed in early 2001, but the fate of this transaction is unclear, because the US manufacturer encountered severe financial difficulties and was acquired by a Singapore-based company.

In 2002–2003 the Egyptian navy received five Tiger high-speed patrol boats armed with Exocet missiles from German navy surplus stocks. They were not renovated before delivery, and therefore the navy is considering having them renovated before commissioning them. Regarding naval armaments, the most significant development in recent years was that the US administration approved of the sale of Harpoon Block II missiles to Egypt. The approval was given despite Israel's strong objections that the missile also possessed the capacities of cruise missiles against ground targets.

➤ Strategic Weapons and Weapons of Mass Destruction

Egypt has two nuclear research reactors: one old of Soviet manufacture, with a capacity of two megawatts; the other, made by Argentina, has a 22-megawatt capacity. The latter is a new reactor, which was completed in 1997. Egypt has signed both the Nuclear Non-Proliferation Treaty (NPT) and the Pelindaba Treaty, which declares all of Africa a nuclear-free zone. Egypt also has an inspection agreement with the International Atomic Energy Agency (IAEA), which governs all its declared nuclear sites. There is no current indication that Egypt has a nuclear weapons program.

Egypt was the first country in the Middle East to use chemical weapons, which occurred in the 1960s in Yemen. Egypt is currently suspected of maintaining chemical and perhaps also biological weapons stores.

Egypt possessed a force of Scud B missiles even before the Yom Kippur War. Since the mid-1980s, Egypt has attempted to reinforce its strategic missile capability, through both the Badr-2000 project (in cooperation with Iraq and Argentina), which was canceled because of American pressure, and in secret projects with North Korea. Cooperation with North Korea gave Egypt the ability to manufacture improved Scud missiles with a longer range, comparable to the North Korean Scud C model. During the 1990s, Egypt managed to double its order-of-battle of Scud launchers, apparently self-produced, and is known to have twenty-four launchers.

➤ Joint Maneuvers

The branches of the Egyptian armed forces have conducted joint maneuvers with many Western countries, particularly in the region. The most prominent military exercise in the past decade was the Bright Star series, held once every two years jointly with the US, many NATO members, and several countries in the region. These maneuvers were systems exercises that stressed planning and the performance of operations involving many participants.

The most recent large scale maneuver was held in 1999. The 2001 exercise was held in the aftermath of the September 11 terrorist strikes, when the US military was preparing to attack in Afghanistan. Nevertheless, 24,000 American soldiers participated in the maneuvers, along with forces from Jordan, Kuwait, and a number of NATO countries. Forty-two thousand Egyptian soldiers took part in the exercise. The planned 2003 maneuvers were canceled because of US operations in Iraq.

➤ Military Aid

Since the Camp David peace accords were signed, the Egyptian armed forces have benefited from extensive US military aid, second only to the aid granted to Israel. In recent years, this aid has amounted to $1.3 billion annually in military aid, and $655 million in civilian aid. Egypt has tried – so far unsuccessfully – to reach an arrangement with the US similar to the arrangement between the US and Israel, under which civilian aid would be gradually reduced, while military aid would increase to some extent.

As in the case of Israel, military aid has affected the nature of Egypt's procurement. Most of Egypt's military procurement is from the US, financed by aid money. Hence also the intended deal for the Moray submarines, which was to be covered by US aid. US foreign military aid does not appear as part of Egypt's declared defense budget, and thanks to this aid, Egypt can afford to spend relatively small sums of its own on procurement from other sources, mostly from the countries of the former Soviet Union. Egypt buys spare parts from these countries and upgrades obsolete Soviet equipment with parts that cannot be acquired from the US.

➤ THE BALANCE OF POWER WITH ISRAEL

The Egyptian armed forces are the only large military force bordering Israel that is equipped with Western arms and that benefits from Western consulting and training. Further significant about the Egyptian buildup, ongoing since the peace treaty with Israel was signed, is its singularity within the Arab world. For various reasons, including the loss of their Soviet patron, the absence of financial resources, military adventures, and sanctions applied to them, most of the other large Arab armed forces in the Middle East have either stagnated or become weaker. It is no surprise, therefore, that the Egyptian armed forces are increasingly regarded as the only Arab entity capable of fighting the Israel Defense Forces (IDF).

This situation poses two questions: Why has Egypt continued to maintain and strengthen such a large military force after signing a peace treaty with Israel that has endured for twenty-five years? To what extent do the Egyptian armed forces pose a threat to Israel?

Egypt has no acknowledged strong enemy in its immediate vicinity. The country has a peace treaty with Israel, and the two Arab countries bordering Egypt, Libya and the Sudan, are militarily very weak. Their orders-of-battle are far smaller than that of

Egypt. It therefore would appear that the Egyptian military buildup is aimed at no other than Israel. Furthermore, reports on the Badr military exercise conducted in 1996 stated that the enemy in the maneuvers was strikingly similar to Israel, and the Egyptian armed forces practiced an outline maneuver resembling an attack on Israel. The identity of the enemy in maneuvers conducted by Egypt since that time has been less defined. The most recent Bright Star exercise in 2001 featured a campaign against armed forces operating according to Soviet military doctrine (this probably referred to Iraq), but it could not be ruled out that some of the maneuvers practiced were theoretically intended against Israel.

As with any country, there are a number of plausible and consistent answers as to why Egypt maintains a large military. The first explanation is the perception of a threat. No statesman in Cairo can ignore the State of Israel and its military power along Egypt's border. The IDF are large armed forces, which even Egyptian military leaders consider to be more powerful than the Egyptian armed forces. Moreover, the Egyptian concept of Israel differs from Israel's view of itself. In Egyptian eyes, Israel is a destabilizing factor – a country with a tendency to use force to solve political problems. Egypt considers Israel's regime to be unstable, with frequent changes in government. Egypt also believes Israel has extremist forces, whose rise to power could increase its aggression. One example of this concern was Egypt's attitude toward the declaration by former Israeli Minister of National Infrastructures Avigdor Lieberman in January 2001 that the IDF was capable of damaging the Aswan dam. In Israel, this statement was dismissed as the episodic eccentricity of an irresponsible person who did not really represent the Israeli government. Egyptians, however, interpreted this statement as a genuine threat by an Israeli political leader. Under such circumstances, any responsible Egyptian leader would feel a need to maintain enough military power to meet the IDF in military combat, should this prove necessary. Against a force like the IDF, a military buildup is necessary, and maneuvers must be conducted in preparation for the ascribed threat.

The second explanation is that the very existence of a strong, modern Egyptian military force creates a deterrent against Israel, and consequently constitutes a factor ensuring the stability of the peace agreement between the two countries. This argument mirrors the Israeli contention that Israel must maintain its military strength as a means of guaranteeing the stability of its peace agreements with Arab countries. Credible deterrence for Egypt requires a large modern military, capable of facing up to the IDF. In addition, a need exists to broadcast deterrent messages to Israel from time to time, in the form of reports about the 1996 Badr maneuver (during Binyamin Netanyahu's term as prime minister). The deployment of the Egyptian armed forces, stationed primarily in the east of the country in the direction of Israel, should also be considered in the context of a deterrent.

The third explanation is Egypt's self-perception. The Egyptians regard themselves as the most important Arab country and the leader of the Arab world, and consequently as the key nation in the primarily Arab Middle East. That is also their historical narrative, which centers on Egypt as a perpetually strong and key power in the Middle East dating back to the pre-Arab period. A country's military power is one of the main elements contributing to its strength, and Egypt cannot afford to be militarily weak in comparison to other Middle Eastern countries.

If Egypt possessed different significant components of strength it could perhaps consider foregoing some element of power, but there is also a gap, perhaps even larger, between Egypt's self-image and its real strength in other areas. Egypt is economically weak, its political influence has diminished, and it has lost its status as a cultural and media center of the Arab world. The gap between the Egyptian self-image and the country's real power is a sensitive source of friction with Israel, since Israel's superiority in military, economic, scientific, and technological fields aggravates Egyptian frustration. This gap encourages competition with Israel and casts Israel as a model for Egypt to imitate. When Israel becomes stronger in some field, and especially when it uses American equipment as a basis for doing so, Egypt seeks to obtain the same status and equipment that Israel has been given.

Like other large organizations, armed forces have an interest in perpetuating and developing themselves. Thus, they constantly strive to find new purposes and adaptations to changing circumstances. The Egyptian military is no different in this respect. On the one hand, it portrays the IDF as a threat against which it prepares and trains in order to justify its budget allocations and buildup. On the other hand, in its search for updated objectives and complementary occupations, the military has invaded other fields that in most countries are purely civilian. As a result, the armed forces are currently a huge economic entity in Egypt. Through the national service administration, founded in 1979, the Egyptian armed forces run dozens of enterprises manufacturing non-military as well as military products, and not just for its own use, but even consumer products for the civilian market. Through its business enterprises, it has begun to assume responsibility for water-management, electricity production, development work, and the manufacturing and supply of medicines. It conducts literacy programs and is almost the sole government agency in the desert regions. A kind of vicious circle has been created, in which an oversized military searches for tasks to occupy itself. The economic success of its endeavors reinforces the military's power, adds to its budget allocation, and enables it to preserve its strength.

The Egyptian armed forces also play an important political role. It is the main bastion of the regime and enjoys a symbiotic relationship with it. The current regime is another link in the succession following the Free Officers coup in 1952, and President Husni Mubarak as well gained leadership through his former military positions. Despite the prominence of the armed forces, however, Egypt did not suffer frequent military coups even in the past, and has usually removed the military from the political arena. The armed forces were kept separate from politics carefully and deliberately, which included special privileges to its officers. As long as the armed forces benefit from being the regime's main source of support, the regime will find it difficult to harm them by cutting their budget and would rather maintain the relationship. The armed forces also serve as a source of employment for many people, in a country where bureaucracy in general is inflated as a means of reducing high unemployment. In this way, the regime buys domestic tranquility at a relatively low price.

Taking this analysis and the environment in which the Egyptian military operates into account, it cannot be concluded that the Egyptian military buildup is exceptional, or that it signals a special motivation to be ready for a specific confrontation with Israel or any other neighboring country. The situation is completely different than the process undergone by the Egyptian armed forces in 1967–1973, which saw a feverish buildup with a timetable designed to bring them as soon as possible to fitness for war against Israel.

The current Egyptian military buildup is relatively moderate, particularly if compared to the buildup rate of the IDF. The procurement of the 880 M1A1 tanks, for example, is spread over fifteen years; the last delivery is not scheduled to become operational before 2008. This amounts to sixty tanks a year, roughly the same rate at which the IDF procures Merkava tanks. Two years ago, the Egyptian air force received the final installment of its order of F-16D planes, and now has no outstanding orders for additional warplanes – whereas Israel is about to receive 102 F-16I planes. The Egyptian military lost its main procurement source, the Soviet Union, in the mid-1970s, and has gradually had to convert most of its equipment to Western armaments, mostly American, while simultaneously preserving the battle readiness of the Soviet armaments still in its service. This is a slow, arduous process; after thirty years, half of the air force's order-of-battle, a third of its armored order-of-battle, and most of its artillery are still based on Soviet weaponry.

The $1.3 billion annual military aid given by the US, usable only for procurement in the US, enables Egypt to maintain its large armed forces, and indeed, a military without an ongoing process for renewing its equipment does not maintain its level of strength; it declines. At the same time, US aid restricts Egyptian flexibility, because the aid must be used to buy US equipment and cannot be converted for other purposes. Egypt thus arms itself mainly with US-made weapon systems, but the US usually fulfills its commitment to maintain Israel's qualitative edge and does not supply Egypt with some of the weapon systems acquired by Israel. For example, Egypt is getting F-16D planes, but not planes equivalent to the F-15Is that Israel has received. Egypt operates AH-64 helicopters and upgrades them to the AH-64D model, but without the Longbow radar given to Israel, which gives those helicopters their special capabilities. The same trend applies to all advanced armament fields. US aid does not cover any spending in local currency, which Egypt has indeed endeavored to cut. Most of the cutback was achieved by reducing the armed forces' manpower and officer corps after the Yom Kippur War, from almost one million soldiers to the 420,000 of today.

As for evaluating the Egyptian threat, the critical question is to what degree the armed forces are truly capable of fighting against the IDF on equal terms in a confrontation situation.

Israel has never enjoyed a quantitative advantage over Egypt; the Egyptian military has always been larger than the IDF, its reserves included. When considering the inventory of what is usually referred to as principal weapon systems (planes, tanks, APCs, artillery, naval vessels), a consistent picture seems to emerge of a narrowing qualitative gap between the two forces. Even now, however, the Israeli air force has more high quality platforms. Israel has approximately 350 F-16 and F-15 aircraft, compared with 200 F-16s in the Egyptian air force, though this is a narrower gap than existed in the early 1980s, when the Israeli air force already had a substantial order-of-battle of advanced warplanes and Egypt had only obsolete Soviet planes. The Egyptian army's order-of-battle in advanced tanks is similar to that of the IDF.

In order to assess, therefore, whether the gap in strength between the IDF and the Egyptian armed forces has narrowed, two questions should be addressed: whether both sides use their weapon systems with equal effectiveness, and whether the bridging of the gap between the inventories or principal weapon systems would be of decisive significance in a possible war between the two armed forces.

For political reasons, most Arab militaries, including the Egyptian armed forces, adopted Eastern bloc weapons and the Soviet military doctrine in the 1950s and 1960s, when they became dependent on Soviet aid. The synthesis proved highly successful. Soviet organization, doctrine, weapon systems features, and maintenance methods proved very suitable for Arab armed forces. To some extent, this reflected resemblances between the Soviet regime and society and the regime and society in Arab countries. All these regimes were autocratic, in which centralized control was very important, dependence on detailed planning by high level officials was absolute, and individuals were allowed no freedom of choice or decision-making flexibility. These factors contributed to the successful absorption and profound internalizing of Soviet methods in those Arab armed forces.

Though the Egyptian military lost Soviet patronage a generation ago, there are no indications that its operating concept has changed since then. The Egyptian military absorbs new weapon systems into its old organization and operates them using the methods to which it is accustomed. The transition from absorbing new weapon systems within an existing organization and doctrine of warfare to developing innovative concepts involving a paradigm shift is a difficult process in any armed forces. In the Egyptian armed forces it is even more difficult, for the same reasons pertaining to society and regime that allowed the easy interface with Soviet methods.

The issue of culture is similarly related. Analysis of cultural differences in order to evaluate military capabilities has acquired a bad name, and is usually a euphemism indicating excessive belittling of the enemy. Nevertheless, it cannot be denied that cultural differences are extremely significant in evaluating military systems, particularly their ability to absorb such profound changes. Discussion of this topic is not scientific, but it cannot be ignored.

Evaluation of the cultural differences between the West and countries like Egypt has not been confined to Westerners. Interest in the subject has recently surged, following the publication of United National Development Programme (UNDP) reports on human development in Arab society.[1] These reports were written by Arab researchers who tried to understand the reasons why the Arab world has lagged behind other regions around the world. Egyptian thinkers have also considered cultural aspects, which they believed were blocking the road to progress and the correct handling of problems. The director of the Al Ahram Research Center, Abdel Moneim Said Aly, stressed the common tendency in the Arab world to disregard one's own mistakes, and blame every problem on an external factor.[2] In a culture where honor confers an advantage and shame a disadvantage, it is difficult to learn and absorb new attitudes and methods, because good learning requires trial and error.

Western military experts who have worked in Egypt and trained its armed forces reported problems that compounded the difficulty in communicating a Western approach.[3] Among their findings:

- Learning skills in Egypt are based on memorizing, and greater value is assigned to remembering details than to understanding methods and principles and using them in analysis.
- The importance of honor prevails. In a competitive situation, even in a class of learners, someone always wins and someone loses – and consequently is humiliated. The tendency is to avoid such situations. It is even more

103

- important to avoid situations in which someone at a junior level appears to know more than someone at a more senior level.

- There is a rigid, inflexible hierarchal structure. Decisions are made from above, and people at a lower level have limited freedom of action. Little is expected from them in terms of planning, improvising, and responding to situations. The absence of a sense of personal responsibility and initiative also affects equipment maintenance, since operating problems are referred to higher levels.

- There is an absence of a culture of cooperation between entities. Cooperation requires trust, and on its deepest level, Egyptian culture dictates distrust of strangers. This attitude makes joint command almost meaningless, because each of the entities making up the joint force will avoid sharing information with other parties, and will view any directive from a different branch with suspicion.

The result of all these factors is that even in a situation of war between two sides that are equal or similar in weapon systems, F-16 versus F-16 or a Merkava tank versus an Abrams tank, the Israeli side will likely be more effective. For example, the Egyptians will operate their F-16s using the same operating concept that they used for operating MiG-21s in the 1970s, as indicated by the format for training and exercises of the present Egyptian air force. The assumption that the two sides' weapon systems are identical is also inaccurate. In a large proportion of cases, Israel uses its highly developed technological infrastructure to alter the weapon systems that it receives and make them more suitable to the specific conditions of the conflicts in which it is likely to participate.

The paradigm for wars between regular armed forces has changed over the past two decades. The most recent reflection of this change and its effect was the Iraq War. In the West, the new paradigm is called the Revolution in Military Affairs (RMA), or Network Centric Warfare. Based on the revolution in information technology, it includes the following key capabilities:

- The use of precision guided weapons, which are able to hit pinpointed targets (down to the level of an individual vehicle). The launchers are largely immune from attack, because the weapons are launched at great distances and the launchers are protected by electronic warfare (EW) devices.

- Information supremacy, which renders the battlefield transparent regardless of visibility, and makes it possible to discover the location of all targets in real time.

- The advent of command and control (C^4I) systems, which make it possible to exploit intelligence information and translate it into information about the targets. This information is immediately transmitted to the attacking level and used to destroy the targets.

This revolution has changed the balance between firepower and maneuver on the battlefield in favor of firepower. A maneuvering force is still necessary, but its goal is to convert the achievements of firepower into attaining the battle goals against a weakened enemy that has lost its fighting effectiveness. The conquest of Iraq by a military force that had undergone this change was a good example. Within a period of less than one month, a force similar in size to the IDF regular army (excluding reserves) effected the collapse of a large army, while suffering less than 200 killed. Previous examples of this type of war included the 1991 Gulf War, the war in Kosovo, and the campaign by the Israeli air force against the Syrian air force and anti-aircraft defenses in 1982. The IDF is the only military in the Middle East that has undergone RMA, absorbed its key capabilities, and benefited from the synergetic advantages of using these capabilities as a system.

Theoretically, the Egyptian armed forces should have encountered the changes in the American paradigm. The teaching and training of Egyptian officers by instructors from the US and other Western countries and the training of Egyptian officers in US military academies were designed to instill the working methods, doctrines, and command habits of the world's most advanced militaries in the Egyptian military. Joint maneuvers with the US such as Bright Star demonstrate how the American system operates. The Egyptian officers have regularly rubbed elbows with American instructors during courses conducted at American instruction institutes such as the US Army War College and the Air War College. An Egyptian force also fought side by side with the US in the 1991 Gulf War, and the significance of the recent 2003 Iraq War from a military perspective is plainly evident. It is therefore clear that the Egyptian military possesses firsthand knowledge of the import of RMA. The question is what the Egyptian military has concluded from its knowledge, what it is doing about it, and whether a war between the IDF and the Egyptian armed forces would be a war between equals, or a contest resembling the war between the Iraqi armed forces and the US armed forces.

There are two main ways of handling a change on the scale of RMA. One is to undergo the same process and acquire the new capabilities. The other is to develop asymmetric responses. Those who are able will attempt to acquire the new capabilities; those who believe they are incapable will concentrate on asymmetric responses. The asymmetric responses are designed to reduce the vulnerability of armed forces lacking RMA capabilities to these capabilities, and transfer the battle to fields in which these capabilities are less effective. One type of asymmetric response is to reduce the signature of the military's force components by forming smaller units, using lower signature weapons, camouflage, concealment, and deception. A switch to terrorism and guerilla warfare is an example of a second type of asymmetric response.

Meanwhile, there are no visible signs that the Egyptian armed forces are developing in either of these possible directions. One result of the development of RMA capabilities is the reduced importance of platforms, accompanied by correspondingly greater importance of the precise armaments they carry and of other force multipliers. In the US, for example, doubts are being voiced regarding the need of the US air force to absorb new advanced platforms. The suspension of such procurement and a focus on advanced armaments is currently proposed. Egypt is still channeling almost all its military buildup to the procurement of new platforms, although it does possess small quantities of precision-guided armaments and an even smaller number of long-range armaments, such as the GBU-15 television-guided gliding bombs. The Egyptian air force has primarily outdated short-range guided weapon systems, such as the Maverick missiles, but procurement of more advanced long-range armaments able to function in any lighting conditions, such as the joint direct attack munitions (JDAM), has proved impossible. Nor does Egypt have long-range air-to-surface missiles, such as the AGM-130 or the Popeye. Egypt's training program and its doctrine of warfare give no indication that these armaments play a key role in its military.

In the intelligence field, Egypt has Beech 1900C electronic surveillance planes and the Hawkeye early warning aircraft, as well as Skyeye unmanned aerial vehicles (UAV) for collecting tactical intelligence. For visual intelligence gathering, Egypt has ordered the Tactical/Theater Airborne Reconnaissance System (TARS)-Podded Reconnaissance System (PRS), but current Egyptian long-range visual intelligence gathering is still inadequate, relying on obsolete photoreconnaissance planes that are unable to provide real time visual coverage of the battlefield. Even more importantly, the Egyptian military is incapable of using its command and control systems to integrate its intelligence gathering, analysis, and operational levels into a system able to identify and pinpoint targets, and allocate precise armaments to them within short response times.

It is therefore clear that at present, the Egyptian armed forces do not possess RMA capabilities. A more difficult question is whether Egypt will decide to institute this change, and whether it will be able to acquire such capabilities in the future. If Egypt does make such a decision, it will likely encounter the following difficult obstacles:

- Presumably the US will persist in its weapons supply policy that is designed to maintain the gap between its own capabilities and those of other Middle Eastern armed forces that could in various circumstances become its enemies. Nor, at this time, is there any indication of erosion in the US commitment to preserve Israel's quality edge. The most recent example of this policy was the US refusal to supply JDAM munitions and Longbow radar systems to Egypt.
- Israel is capable of overcoming problems caused by US weapons sales policy through the use of its independent development and manufacturing capabilities. Egypt has no such capabilities.
- The cost of a paradigm change in the Egyptian armed forces would be enormous. Even if US aid continues, Egypt will find it difficult to shoulder this cost.

In the last analysis, the main obstacle facing potential Egyptian RMA consists of the same difficulties that have hampered the process of change in the Egyptian armed forces for the past three decades. Armed forces adopt a new doctrine when it becomes clearly evident that the current doctrine does not fulfill their needs and is no longer suitable. The question is whether the Egyptian military truly feels that its doctrine fails to meet its needs. The Egyptian military's awareness of the existence of RMA does not necessarily mean that its leadership believes that this doctrine is suitable for it. It is possible that the lack of indications that Egypt is dealing with this field means that the Egyptian military leadership has already decided against moving in this direction.

The Egyptian military regards its main role as defensive and as constituting a deterrent to Israel. In accordance with this philosophy, it seems to feel fairly confident in its strength and believes that the quality and quantity of its equipment provides an adequate counterweight to Israel. The Egyptian military will no doubt aspire to upgrade this equipment, but so far it has evinced no serious impulse to change its military doctrine.

➤ Is a Military Confrontation Realistic?

The Israeli–Egyptian peace agreement has proven its stability by enduring difficult tests, including the war in Lebanon and two rounds of a violent conflict between Israel and the Palestinians. The peace agreement has also affected the deployment of military forces by Egypt and Israel. Due to the restrictions imposed by the agreement, most of the Egyptian army is stationed west of the Suez Canal. There is a buffer zone some 300 kilometers wide where a multinational peacekeeping force is stationed between the two armies. Egypt belongs to a bloc of countries aligned with the West and in particular the US, and benefits from large amounts of aid from the US and other Western countries. A military conflict with Israel is possible only if a basic change occurs in this strategic situation, leading to a change in Egypt's strategic orientation.

For such a change to occur it would not be enough if Egypt's head of state were replaced. Egyptian President Anwar Sadat was murdered and was replaced by Husni Mubarak, but Egypt's basic strategy did not shift. Rather, a change of this sort would involve a decision to secede from the bloc of moderate countries enjoying close relations with the US. The dissolution of the Soviet Union, which left only one superpower in the world, has rendered a decision of this nature even more difficult. There is no alternative to what the US supplies Egypt. In this reality, the only scenario that might give rise to this kind of strategic change is the replacement of the current regime by an extremist Islamic regime of the Iranian type, which sees no need for close relations with any superpower.

Without addressing the question how likely the assumption of power by a revolutionary Islamic regime is, it is clear that this would also affect Egypt's military power. It would be wrong to assume that such a change in Egypt would leave its power intact and merely channel it in different directions, just as this assumption was proved wrong in the case of the Khomeini revolution in Iran, due to Iran's loss of its principal weapons supplier – the US. The effect could prove even greater in the case of Egypt; while Iran relied on the US as a weapons supplier, it had its own sources of financing. Egypt, however, relies on the US for both money and weaponry. The conclusion is that a decision by Egypt to enter a military confrontation with Israel would lead to a decline in its military power.

In a military confrontation against Israel, Egypt would have to cross the distance between the Suez Canal and the Israeli border. The army that would have to accomplish this is a mechanized and armored army prepared for Six Day War and Yom Kippur War types of conflict. Its ability to fight against RMA capabilities, however, is extremely limited. The topography and the terrain of the Sinai Peninsula make it an ideal region for a force conducting RMA warfare. Mechanized and armored forces in motion cannot be concealed there, and their rate of attrition would be high.

Thus, Egypt's continuous military buildup in recent decades has indeed given it capabilities that it did not previously possess. The air force has an order-of-battle of attack warplanes with a range enabling it to attack targets within Israel, as well as planes with better fighting capabilities in air-to-air combat. The land forces are almost completely mechanized and possess the capability for in-depth attack. The navy is a large, highly developed fleet of surface ships. All these capabilities could play a key role in a war like the Yom Kippur War. In a conflict of the type expected in the early twenty-first century, however, they will be relatively unimportant. Furthermore, even in areas where Egypt has built up its strength, the IDF buildup has been at least as massive, and more so in most areas. The gap between the IDF and the Egyptian armed forces in the most important aspects has not narrowed; it has widened.

The Egyptian military and political leaderships are aware of this gap. It was no coincidence that at the end of 2000, when voices in the Arab world called for Egyptian military intervention in the conflict between Israel and the Palestinians, Mubarak answered harshly, declaring that he had no intention of dragging Egypt into a military conflict with Israel. He certainly had good political reason for opposing such a venture, since he realized the price a change in Egypt's political orientation would exact. It can also be concluded, however, that awareness of the military balance likewise played a role in Mubarak's thinking.

Notes

1   *Arab Human Development Report 2002*, UNDP, 2002, and *Arab Human Development Report 2003*, UNDP, 2003.
2   Abdel Moneim Said Aly, "Confronting the Conundrum," *Al Ahram Weekly*, October 31–November 6, 2002. (This was the first in a series of eleven articles on the same subject published between October 2002 and February 2003).
3   Norvell de Atkine, "Why Arab Armies Lose Wars," *Middle East Review of International Affairs (MERIA)* 4, no. 1 (2000).

CHAPTER ▶▶▶

7

Israel at a Strategic Crossroads

SHAI FELDMAN

On June 6, 2004, the Israeli government adopted Prime Minister Ariel Sharon's disengagement plan. The plan calls for Israel's "withdrawal from the Gaza Strip and from the northern part of Samaria." The withdrawal from Gaza is to include all military installations – with the exception of the area adjacent to the border between Gaza and Egypt – and all Israeli settlements. The plan also states categorically that the "State of Israel will withdraw from northern Samaria . . . and will redeploy outside the evacuated area. . . . Once the move has been completed, there will be no permanent Israeli military presence in the area."

The Israeli government's decision apparently reflected the conclusion reached by Prime Minister Sharon more than a year and half earlier – a conclusion shared by a clear majority of the Israeli public – to the effect that Israel must separate itself from the Palestinians and end its control over large areas of Gaza and the West Bank. In turn, this conclusion seems to have emerged from an appreciation of the paradox of Israel's relationship to its regional environment.

At the regional level, Israel has reached a level of primacy not unlike the preponderance achieved by the United States at the global level in the aftermath of the Cold War. At the same time, it has become increasingly clear that Israel's supremacy in the region will not compel the Palestinians to lay down their arms and end terrorist attacks against Israel. Moreover, Israelis have become increasingly aware that they cannot compel the Palestinians to accept a permanent resolution to the conflict under terms inferior to those brokered by President Clinton in 2000. Finally and most important was the recognition that strategic primacy provides no remedy for the growing demographic threat facing Israel, and that given present population trends Jews would soon lose their majority in the area between the Mediterranean Sea and the Jordan River.

In 2003–2004, this disparity between supremacy and victory seems to have led Prime Minister Sharon to reverse his longstanding approach to the Arab–Israeli conflict by making two important decisions: one operational, one grand-strategic. The first involved supplementing Israel's offensive counterinsurgency strategy with a strong component of static defense, specifically, the construction of a security fence between Israel and the West Bank. At the grand-strategic level, the decision was to launch a process of separating Israel from the Palestinians of Gaza and the West Bank through Israeli withdrawal.

How far and at what pace Israel would be able to move toward overall separation from the Palestinians remained in mid-2004 an open question. Just as the precise demarcation of the fence has become subject to considerable conflicting pressures, the attempt to separate Israel from the Palestinians became a subject of deep disagreement among Israel's leadership and political elites.

➤ ISRAEL'S INTERNATIONAL STANDING

According to almost any indicator of national power, by the early 2000s Israel has achieved a position of strategic primacy in the Middle East. In the effectiveness of its conventional military forces, it is unchallenged not only by any of its neighbors but also by any possible coalition of Arab states. It is the only state in the region perceived by its neighbors as possessing a full array of non-conventional weapons – nuclear, chemical, and biological – as well as the ability to deliver these weapons to every corner of the region.

In macro-economics, Israel's GDP is larger than all of its neighbors' combined. Its GDP per capita is now in the same league as a number of West European states and by the year 2004 its GDP per capita would have likely equaled that of the United Kingdom if it were not for the effects of the violence launched by the Palestinians in September 2000. While in 1980 Israel's GDP per capita was but 40 percent of that of Saudi Arabia, by 2000 the tables had turned: Saudi Arabia's GDP per capita was only 45 percent of Israel's. In addition, Israel's foreign currency reserves are now, at over $26 billion, their highest level ever, more than four times higher than their levels in the 1970s and 1980s.

In science and technology, Israel is widely recognized as one of the world's most advanced states, sometimes dubbed the region's Silicon Valley. In its neighbors' eyes, its technological edge is very difficult to match, since it rests on a "culture of technology" in which Israeli society is said to be embedded.

In addition, Israel enjoys a de facto alliance with the United States, the only remaining superpower. This alliance appears quite robust, resting on shared values, a common commitment to democratic government, an array of complementary strategic interests, and the political influence of the American Jewish community and the Christian Right. The role of these sectors is unparalleled in Western democracies: it is based on a political culture that regards interest groups as a legitimate part of the democratic process and on campaign finance laws that allow such groups to influence elected officials. Following the September 11, 2001 terror attacks, the near-formal US-Israel strategic alliance received a further important boost. In the past, two countries have sparred over the IDF's strategy of preemptive offense as well as over Israel's efforts to deter terrorist attacks through punitive measures. But the National Security Strategy adopted by the United States in late 2002, emphasizing the need for preemptive and preventive action, brought the two countries' defense policies increasingly into sync.

Parallel to Israel's growing strength, its Arab neighbors have considerably weakened over the past two decades. In the late 1980s, a number of Arab states lost their principal strategic ally, the Soviet Union, and have not been able to find a substitute for the assistance that Moscow provided them in previous years. Also, Arab unity, manifested impressively in the early 1970s in the oil embargo imposed against the West, has long since evaporated. Since then, Arab foreign and defense policies have become increasingly guided by particular national interests, rather than by pan-Arab considerations. This was especially pronounced in 1990–1991, when Iraq invaded Kuwait. The invasion of one Arab state by another led to the Gulf War, in which Iraq was defeated by a coalition that included the forces of two important Arab states: Syria and Egypt. Furthermore, during the 1990s the Arab League has become increasingly non-functional. Indeed, by the first half of 2004, merely convening its leaders has presented a major challenge. Thus, most Arab states have lost their capacity to shape the Middle East agenda, and instead, three non-Arab states have assumed this leading role: Israel, Turkey, and Iran.

While these facets of Israel's evolving preeminence developed gradually during the 1980s and were heightened in the 1990s, they received a further boost as a consequence of the 2003 Iraq War. With the Iraqi military destroyed, the danger of an "eastern front" against Israel was put to rest. This was likely to remain the case even after the United States reduces its military presence in the region from its current high level. The apparent absence of weapons of mass destruction in Iraq, Libya's decision to renounce its WMD and abandon its related programs, and the international pressures to persuade Iran to halt its nuclear program combined to diminish significantly the overall WMD threat facing Israel.

In addition to these strategic gains and almost equally important, by the early 2000s Israel's foreign and international economic policy scored impressive achievements. First, Israel has succeeded in preserving its signed peace treaties with Egypt and Jordan. The former has survived more than twenty-five years, despite many trials and tribulations, and relies on a distribution of power which is extremely favorable to Israel. While Israel's peace with Jordan is less than ten years old, it seems at least as strong, resting on complementary strategic interests. This has made Jordanian-Israeli relations relatively immune to negative developments, despite Jordan's Palestinian majority.

Second, in sharp contrast to the near-total isolation that Israel suffered in the 1970s, it is now completely integrated in the international community. In addition to developing relations with Russia and the Eastern European states and to rebuilding its ties with most black African states, it has forged new relations with the giants of Northeast Asia: China, Japan, and South Korea. While Israel's improving defense ties with China suffered a serious setback in mid-2000 due to Israel's cancellation of the agreed sale of Phalcon early warning and control aircraft to China, over recent years it has built an important strategic alliance with Turkey and with India. In both cases, what began with modest trade and cooperation in the defense-industrial sector later developed into a full array of economic interactions, from tourism to trade in civilian products.

More generally, by the 1990s Israel has become completely integrated into the global economy. The products of its hi-tech sector are marketed almost everywhere, from Latin America to the Far East. Similarly, Israeli firms are involved in a number of major energy projects from Egypt to Turkmenistan. In Western Europe, Israel's successes are more modest. But whereas most media reporting focused on criticism leveled at Israel by the continent's foreign ministries and liberal press, ignored was the considerable sympathy for Israel's predicaments among the defense communities of the same West European states. Also ignored was the web of commercial relations tying these states to Israel, and indeed, Europe remains Israel's largest trading partner.

Another dimension to Israel's strength in foreign policy involves its relative success in marketing its basic narrative regarding the Palestinian–Israeli conflict, namely, that the responsibility for the failure to reach a negotiated end to the Palestinian-Israeli dispute rests primarily with the Palestinians in general, and with PLO leader Yasir Arafat in particular. Arafat is cast as having rejected the hand extended to him in 2000 despite then Israeli Prime Minister Barak's apparent willingness to break every taboo in order to present the Palestinians with a generous diplomatic offer.

In the United States, Israel's success in de-legitimizing Arafat was echoed by President Bush's conviction, expressed in his June 24, 2002 speech, that Arafat's replacement by a new leadership "not tainted in terror" was a prerequisite to any real progress in the peace process. In Europe, Israel's campaign led to increased sensitivity if not anger at the level of corruption characterizing the Palestinian top leadership. The result was conditioning continued European assistance on Palestinian financial openness open and on nominating a finance minister committed to the principles of transparency and accountability.

Ironically, by 2003–2004, it seemed that in the international arena the main problem facing Israel was the consequence of its close alliance with the United States. In other words, in the eyes of many in Europe, and certainly among Arab governments and the international public, Israel and the United States have become indistinguishable, thus inextricably linked with one another's misfortunes. As a result, the United States was held increasingly accountable to all Israeli acts of misconduct. Many Europeans and Arabs alike insisted that Israel could not have continued its occupation, its expansion of settlement activities, and its punitive attacks on Palestinians without Washington's consent. Similarly, Israel became associated with the American occupation of Iraq. And as both countries' counterterrorism efforts have often failed to distinguish between combatants and non-combatants, each has been considered guilty of the other's mistakes.

In Europe in particular, this problem was compounded by the widespread contempt that developed in the continent toward US President George W. Bush and Israeli Prime Minister Ariel Sharon. While the basis for the emotional reaction to the two leaders differed – the first was propelled by the Bush administration's perceived reliance on unilateralism and by the administration's own contempt for Europe, while the latter rested mainly on Sharon's past record, particularly as the architect of Israel's 1982 invasion of Lebanon – the close relations forged between the two leaders seemed to elicit venomous reactions in Europe's media and among its political and intellectual elites.

➤ THE PALESTINIAN CHALLENGE

These complications notwithstanding, by 2003–2004 the main challenge facing Israel was that its regional strategic supremacy could not be translated into an ability to impose a solution on the Palestinian-Israeli conflict. True, from Israel's perspective the supremacy it had achieved has had an important impact on the conflict: by providing Israel with effective deterrence at the state-to-state level, its primacy has helped prevent the escalation of the violence into a regional war. Thus, for example, the possibility that Egypt might be impelled to violate its peace treaty with Israel was dismissed by President Mubarak already in 2000–2001. Egypt, he made clear, was not about to get involved "at the expense of the last Egyptian soldier." Similarly, fears of escalation along Israel's northern border did not materialize, as Syria's leaders were highly attuned to the Syrian–Israeli balance of power. As a result, any Palestinian hopes that important Arab states would come to their rescue and redress the Palestinian–Israeli imbalance by embroiling themselves in another Arab–Israeli war were proven empty.

Furthermore, the Palestinians have for their part failed to gain any important achievements in their violent struggle against Israel. Thus, despite more than three and a half years of relentless terrorism, resulting in a thousand Israelis dead, thousands of wounded, considerable damage to the Israeli economy, and even greater damage to Israel's standing abroad, the Palestinians have little to show for their own enormous losses in human life and the near total destruction of their economy. Consequently, by mid-2004 they appeared no nearer to their goals than they were in late 2000, in the aftermath of the collapse of Palestinian-Israel negotiations orchestrated by President Clinton.

In no small measure, this Palestinian failure resulted from the impressive stamina manifested by the Israeli society, which seemed to rest on a wide consensus that although there was enough blame to be spread around, primary responsibility for the ongoing conflict rests with the Palestinians. Thus, most Israelis remained convinced that in late 2000 their leaders made substantial efforts to bring about a negotiated end to the Palestinian-Israeli dispute.

Yet even the combination of Israel's strategic supremacy and the resilience of its society were not agents that could compel the Palestinians to end the conflict. Nor did such primacy suffice to coerce them to lay down their arms and cease terrorist attacks against Israel. Even more at issue, however, was the irrelevance of strategic primacy to the ever-growing demographic threat. With Jews already comprising only a slight majority between the Mediterranean and the Jordan River, given existing demographic trends they could expect to become a numerical minority in the near future. The particular danger, aside from the threat of being at a numerical disadvantage, was that at that point the Palestinians would abandon their long-standing quest for independence based on a two state solution and instead demand full participatory rights within a single state. Israel would then have to choose between denying this demand – thus compromising its democratic character – and accepting it, thus abandoning its commitment to Jewish statehood.

The perceived absence of a Palestinian partner for a negotiated end to the conflict and the disparity between strategic primacy and the demographic challenge seem to have led Prime Minister Sharon in 2003–2004 to make two important decisions: to accelerate the construction of a security fence dividing Israel from the West Bank and to advocate the disengagement of Israel from Gaza as well as from a northern part of the West Bank. The latter proposal was also propelled by Sharon's growing concern that if Israel would not take the strategic initiative, others would, and Israel would then be relegated to the defensive role of constantly battling the initiatives not to its liking.

> ### Israel's Unilateral Measures

> ## The Security Fence

Erecting a barrier between Israel and the West Bank was opposed by Sharon for many years, lest it be interpreted as Israel's willingness to abandon its claim to the West Bank. But in 2002, the frequent terror attacks in Israel's large metropolitan centers prompted a broad public demand that Israel's offensive efforts to destroy the terrorist infrastructures in the territories be supplemented by strong defensive measures. As a result, Sharon felt he had no choice but to accede to the idea of a security fence and advocate its construction at an accelerated pace. The proposed fence system was to be similar to that which Israel erected in the early 1990s around the Gaza Strip and along Israel's border with Lebanon following the IDF withdrawal in May 2000.

The decision to build a fence was opposed by large parts of the Israeli right wing and especially by Israelis residing in West Bank settlements. Their opposition stemmed primarily from concern that the fence would initiate a process of their separation from the majority of Israelis, those residing within Israel's pre-1967 boundaries. Moreover, they interpreted the fence as signaling that their security was a secondary consideration, as enormous financial resources would be invested in making Israelis within the pre-1967 lines more secure. In turn, if Israelis inside the pre-1967 boundaries would feel more secure, their willingness to continue investing in the security and defense of the settlements might well diminish. In addition, the Israeli right and the settler community feared that as the fence would become a more effective barrier to reaching Israel's large metropolitan areas, the terrorists' energies would be channeled almost exclusively against those residing in the territories, thus making their lives intolerable.

Once objections were overruled and the Israeli government decided to construct the fence, the opponents of the enterprise redirected their efforts to determining the fence's demarcation (figure 7.1). They argued that its construction should be kept well east of the 1967 line, lest it be interpreted as Israel's willingness to withdraw to 1967 lines in a future permanent status agreement. They further argued that it was unreasonable to expect the vast majority of Israelis residing in the large settlement blocs east of the 1967 lines to remain unprotected.

Thus, the demarcation of the fence became a central focus of public debate in Israel during 2003, as the pressures from the right met important counter-currents. Many in Israel's defense community argued that the construction of the fence east of the pre-1967 lines would be self-defeating, as it would imply that more Palestinians would find themselves residing west of the fence and thus within easy reach of Israel's population centers. Even more important, considerable pressure began to mount abroad, as the Palestinians succeeded in persuading key governments as well as a significant part of international public opinion that a fence east of the 1967 lines represented Israel's attempt to predetermine the outcome of any future permanent status resolutions. Additional external pressures centered on the humanitarian implications of the fence, as its eastern demarcation separated many Palestinians from their agricultural fields as well as from their schools and hospitals, thus compounding the hardships they were already experiencing.

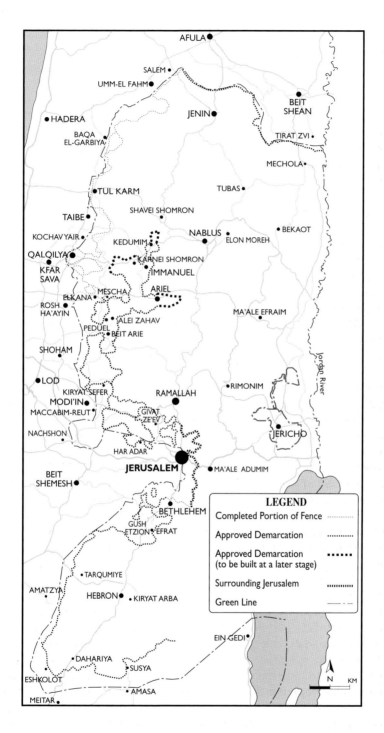

Figure 7.1 Demarcation of the Security Fence
Based on the map that appeared on the Ministry of Defense website, December 2003

As the construction of the fence became a hotly contested issue, it was necessarily thrust on the Bush administration's agenda, especially since the United States was already held accountable for Israeli action. Consequently, the precise demarcation of the fence became a focus of intense US-Israeli consultations, with National Security Advisor Condoleezza Rice becoming personally involved. On June 29, 2003, during her only official trip to Israel to date, Rice met with the Israeli cabinet to discuss the administration's concerns about the fence.

In early 2004, the Palestinian Authority and a number of Arab states brought the issue to the International Court in the Hague, arguing that the fence had become a "crime against humanity." To the Israeli government's surprise and dismay, the subject also became a source of considerable tension with Jordan. The Hashemite Kingdom took strong exception to the construction of the fence, fearing that the resulting dislocations would create added demographic pressures from the West Bank to Jordan.

Under these enormous cross-pressures, the Israeli government began to vacillate in its decisions about the demarcation of the fence. In one location this occurred after the fence was already fully constructed, thus requiring the segment to be demolished. Observing the indecision, one Israeli journalist humorously suggested that his government might consider putting the fence on wheels.

➤ Initiating Separation

As the debate regarding the demarcation of the security fence continued in tandem with its construction, Prime Minister Sharon apparently experienced a much deeper change of heart about Israel's relations with the Palestinians. In other words, he seemed to embrace what he had rejected for nearly thirty years: an end of Israel's occupation of the Palestinians and, perforce, a separation between the two communities.

Although Sharon provided hints of a possible strategic about face a year earlier, in a December 2003 speech at the Herzliya Conference he argued that Palestinians should be allowed to create a state, and barring a negotiated plan, Israel should implement a plan of unilateral disengagement. Therefore, he pointed out, Israel should withdraw from the Gaza Strip as well as from parts of the West Bank. Such a pullout would thus require the "relocation" of all Israeli settlements in the Gaza Strip as well as some settlements in the West Bank. He further stressed that such relocation was necessary to allow the emerging Palestinian state to enjoy "territorial contiguity."

Sharon's apparent broad shift is not easy to explain. One possibility is that it reflects his internalization of the paradox of Israeli power: that strategic supremacy does not suffice to end the conflict with the Palestinians and to quash the demographic threat. This rationale figures prominently in the views expressed by the strongest proponent of unilateral separation within the Likud, Vice Premier and Minister of Trade and Commerce Ehud Olmert. Yet Sharon has never associated himself with the comprehensive vision that Olmert has outlined in Israel as well as in Washington since late 2003.

Another possibility is that Sharon's plan was a natural consequence of successful Israeli efforts to persuade the international community that a negotiated resolution of the conflict was impossible as long as Arafat remained in place. Hence the inevitable challenge to the Israeli government, "so what do you propose?" Indeed, even within Israel, Sharon was increasingly asked whether the absence of a negotiating partner meant that Israel was relegated to the role of a passive observer in the face of the negative demographic trends. Moreover, in 2003 there were indications that Sharon became increasingly concerned that by not providing a proactive vision of its own, Israel was creating a vacuum that might soon be filled by alternative plans for dealing with the Palestinian–Israeli conflict – plans he feared might be extremely harmful to Israel. One salient example was the Geneva initiative negotiated that year by a group of Israelis led by opposition figure Yossi Beilin and by a group of Palestinians led by Yasir Abed Rabu.

A less generous explanation was that by the end of 2003 Sharon became absorbed by the need to deflect attention away from the police investigations then underway regarding different aspects of his family's finances. The assumption was that the desire to avoid derailing a major strategic change would figure highly in the considerations of Israel's Attorney General on whether to indict an acting prime minister.

In the Middle East and in Europe, Sharon's pronouncements were initially met by considerable skepticism. These doubts centered on Sharon's seriousness about implementing his proposal as well as the extent to which he was truly prepared to make the sacrifices that real separation entails. His statements were largely viewed as tactical ploys aimed at repelling other initiatives and as a desperate attempt to divert attention from his legal predicaments.

From the outset Sharon attached critical importance to Washington's reaction to his proposal. The administration's main concern about the plan seems to have been how it might contradict the vision articulated by President Bush in his June 24, 2002 speech. Specifically, implementation of the plan might lead to chaos in Gaza and the West Bank, thus destroying any hope of institution building in the Palestinian Authority and of combating terrorism effectively by restructuring the Palestinians' security services. For its part, the State Department was worried about the degree to which Sharon's plan might end any hope of reviving the stalled roadmap to Middle East peace announced in the aftermath of the Iraq War. This, however, appeared to be of lesser concern in the White House, which apparently shared Sharon's assessment that implementation of the roadmap was impossible as long as Arafat remained at the Palestinian helm.

On a more fundamental level the administration's initial concerns about the disengagement plan seem also to have centered on the wider implications of Israel's withdrawal. With the situation in Iraq viewed as increasingly complicated and with the administration finding itself under growing heat as to its intelligence prior to the September 11 attacks and its subsequent decision to invade Iraq, it feared that Israel's unilateral withdrawal might create another mess to which the United States would be held accountable.

Washington's reservations, however, were soon replaced by a decision to help Sharon implement his plan. The reasoning behind the switch seems to be have been that even if falling far short of providing a total solution to the conflict and to Israel's security and demographic challenges, any removal of Israeli settlements will begin to reverse the dynamics of the conflict. Given Israeli domestic political realities, Sharon was seen as the only Israeli leader capable of unweaving Israel's occupation of the territories and of beginning the process of dismantling Israeli settlements.

➤ MARKETING THE DISENGAGEMENT PLAN

Parallel to the attempt by Israel's National Security Council, headed by Major General Giora Eiland, to translate Sharon's idea into an operational plan, the Israeli prime minister initiated an effort to gain political approval for his proposal. In facing this challenge, he was confronted by sharp components of Israeli domestic politics and the decision-making apparatus. First, every attempt to gauge Israeli public attitudes on the issues involved indicated that a clear majority supported Israel's separation from the Palestinians – from 55 to 72 percent, depending on the manner in which the question was worded. In contrast, however, it soon became apparent that there was strong opposition to the plan among all party organs: the Likud government ministers, the party's faction in the Knesset, and its Central Committee – a body numbering almost 3000 party activists.

Second, because of the weakness of the Labor party and its leadership, Sharon apparently believed that he could not count on Labor support in the same manner that Menachem Begin had relied on Labor in 1978 to win Knesset approval for the Camp David Accords. Therefore he opted not to bypass the Likud Party organs and bring his plan directly to a Knesset vote. Skeptical of the requisite support in the Knesset, he also refrained from raising his plan in a national referendum since the move itself and the precise wording of the question to be placed in such a referendum would have required the Knesset's approval.

Third, since the separation plan's security ramifications seemed to feature highly in the public mind, support for the proposal among the top ranks of Israel's defense community also became important. Yet given their professional uniforms, these defense officials initially adopted a very cautious stance, dwelling on the serious risks involved as well as the plan's many possible pitfalls. Thus, they emphasized the tactical and operational problems entailed, while de-emphasizing the plan's potential grand-strategic advantages. In turn, the risks and security challenges were seized by opponents of the plan to build their case.

Confronted by these realities, in early 2004 Sharon adopted a three-pronged strategy for winning support for the plan. First, he scaled down the plan from the all-encompassing program that he presented in December, to a more narrow project focusing on a complete withdrawal from Gaza but only a token pullout from the West Bank – the latter requiring the "relocation" of only four sparsely populated settlements. This decision was based on public opinion surveys indicating that the continued control of Gaza enjoyed least public support. Thus, Gaza seemed the point of least resistance to the plan, and it was there that opposition to Sharon's plan seemed easier to break. Second, Sharon decided to submit his plan to a vote by all registered members of Likud, assuming that opposition to a pullout was likely to be weaker among Likud's registered members than among the party's higher organs.

Finally, Sharon decided to offset objections to the plan by winning the endorsement of the Bush administration and its public commitment to back key Israeli positions in future efforts to resolve the Palestinian-Israeli conflict. Consequently, what began as a US–Israeli mutual exploration of Sharon's ideas was soon replaced by lengthy negotiations focusing on the precise text of the assurances that President Bush would provide Israel to help win domestic support for Sharon's plan.

To conduct these talks, President Bush nominated a joint NSC–State Department team that included Deputy National Security Adviser Stephen Hadley, Special Assistant for Middle East Affairs Elliot Abrams, and Assistant Secretary of State William Burns. The Israeli team named by Sharon included his Chef de Cabinet, Dov Weisglass, the NSC's Eiland, and Foreign Policy Adviser Shalom Turjeman. During six negotiating sessions, the two teams worked out a set of understandings which were finalized during Sharon's visit to Washington in mid-April 2004, in the framework of letters provided by President Bush to Sharon as well as from Sharon to the president, and a separate letter delivered by Weisglass to National Security Adviser Condoleezza Rice.

The letter provided by President Bush to Sharon included a number of assurances regarding a future resolution of the Palestinian-Israeli conflict. Most important, it stated that the United States recognizes the demographic changes that have occurred over the years in the West Bank, implying that an agreement ending the conflict would not require Israel to abandon the large settlement blocs that it had built close to the 1967 lines and around Jerusalem. It also stated that a resolution of the conflict would allow Jews to return to Israel and Palestinians to return to Palestine, but not a return of Palestinians to Israel.

To improve the prospects of winning support for his plan among Likud members, Sharon promoted the assurances provided in the Bush letter as a significant development in the US position regarding the contours of a permanent status Palestinian-Israeli agreement. And in order to help Sharon's prospects, the White House refrained from denying the exaggerated interpretations of these assurances, which Sharon and his aides successfully spun in the Israeli media. In turn, the Arab media and much of the European press adopted these interpretations almost without question, thus creating the impression that the Bush administration had significantly deviated from past US positions and that it had allied itself squarely with the most extreme Israeli demands in future permanent status talks. Not surprisingly, this interpretation resulted in widespread condemnations if not expressions of horror in Europe and the Middle East.

A careful reading of the Bush letter reveals that its text does not deviate significantly from the parameters of a permanent status Israeli–Palestinian agreement articulated by President Clinton before he left office. The letter also points out that it is not intended to prejudge future negotiations and that the issues mentioned would need to be resolved in the framework of eventual Palestinian–Israeli talks. Thus, what was meant to help win Israeli domestic support for Sharon's separation scheme was magnified way beyond its intrinsic value, with regional and international repercussions.

Yet even the hyped commitments provided by President Bush proved insufficient to persuade registered Likud members to support Sharon's separation plan. On May 2, 2004, they rejected their leader's proposal by a ratio of 60:40. This slap-in-the-face can be explained by several factors. First, from the beginning, there had been a clear asymmetry between the plan's proponents and its opponents: while the former were initially more numerous even among Likud party members, the opponents of the plan were more motivated, more committed, more mobilized, and better organized. Thus, they were better prepared to wage a campaign to defeat the prime minister's proposal.

Second, Sharon never formulated an argument for the plan and he refrained from launching a comprehensive campaign to win its approval. Instead, he seemed to rely on the premise that his party's members would not dare abandon their leader. In turn, the same party members seemed to interpret the absence of a campaign and the failure to construct an argument to support the plan as indicating that they were being taken for granted. In contrast, the opponents of unilateral disengagement – the most important of which were the leaders of the Council of Settlements – stepped into the void and ran an effective campaign to persuade Likud members that the consequences of Sharon's plan would be catastrophic.

Part of Sharon's mistake was to invest too much in the effort to obtain the Bush administration's approval and its aforementioned assurances. This angered his government ministers who took exception to the fact that Sharon had revealed his plan in Washington before he showed it to them. Given the small size of his staff and his limited time and political capital, it was probably also a mistake for Sharon to divert energy away from managing the campaign within Likud to gaining Washington's blessing. In the end, it was not clear whether the Bush assurances made any difference to Likud party members who actually participated in the referendum – some 75,000 out of the 193,000 eligible.

➤ A Dysfunctional System?

While the outcome of the Likud vote can be largely explained by a number of tactical-political errors made by Prime Minister Sharon, the manner in which the internal battle over the separation plan unfolded has raised deeper questions about the decision-making mechanisms in Israel and even greater doubts over whether its political system will ever allow an end to the occupation of the Palestinians. In the words of a senior Israeli journalist, the question was whether Israel's leadership would ever be able to end the 1967 war.

The dysfunctionality of the Israeli national decision-making system appears to include the following: first, the present configuration of Israeli politics can prevent the translation of the majority's preferences to action. Most of the responsibility for this state of affairs lies squarely with the majority: democracies provide mechanisms, both parliamentary and extra-parliamentary, that measure not only the distribution of opinions but also their intensity. As the battle over the separation plan demonstrates, the majority will be defeated as long as they do not match the degree of commitment, mobilization, and organization demonstrated by the minority.

Second, Israel's particular brand of democracy – a parliamentary system based on proportional elections – continues to produce highly fractured and dysfunctional coalitions that require the prime minister to devote much of his time, energy, and political capital to coalition maintenance. It also provides considerable extortion potential to the smallest of the parties comprising the governing coalition. Third, the staffs available to Israeli prime ministers are much too small to allow the efficient management of a complex political campaign. Prime Minister Sharon's staff proved no exception. With almost everything resting on a single Chef de Cabinet, the prime minister's office proved incapable of performing more than one task or of dealing with more than one crisis at a time.

Fourth, Israel's obsession with security provides the IDF and the security services a disproportional say in national decision-making. While at the formal level Israel can be considered a model democracy as far as civil–military relations are concerned, in practice the IDF and the security services enjoy huge influence in forming the political echelon's understanding of the "reality" against which policy needs to be carved. For instance, before the government began to debate Sharon's separation plan on May 30, revised in light of the Likud vote defeat, briefings were provided by the IDF's Chief of Staff, the Director of Military Intelligence, and the Head of the General Security Services. In contrast, the Director General of the Foreign Ministry and the director of the Foreign Ministry's Intelligence and Research Branch were not asked to present the political-diplomatic backdrop to the plan. Similarly, in its more than five years of existence, Israel's National Security Council has failed to acquire the necessary "gravitas" – and successive Israeli prime ministers have failed to make sure that the NSC will enjoy the requisite standing – that might have allowed it to play a more central role in the national decision-making process.

One consequence of this asymmetry is that when weighing strategic choices, emphasis is inevitably placed on the narrow security and military dimensions of the options at the expense of wider strategic considerations. Worse still, the disproportionate importance given to the military results in an emphasis on the risks entailed in the choices considered, at the expense of the opportunities which the very same options may open. And yet the most important aspect of this dysfunctionality is that strategic choices are considered without first defining and articulating the basic purposes of the state. As a result, the reference points against which the ramifications of these options should be measured are left ambiguous. This, sadly, was reflected clearly in the debate over Sharon's plan.

➤ AN INTERIM ASSESSMENT

By mid-2004, some uncertainties about Sharon's proposals were lifted while others remained. First, it has become increasingly clear that Sharon was serious and determined to implement his proposal. Indeed, in early June he fired two government ministers who were staunch opponents of his plan in order to gain a technical majority in his government to approve the proposal. This left his government with a minority of fifty-nine seats in the 120–seat Knesset.

Second, it became evident that the crux was not how far Sharon was willing to go in implementing his plan. Indeed, international leaders concluded that even a partial implementation of Sharon's limited plan would likely change the dynamics of Israeli–Palestinian relations. And it became increasingly clear that the process of Israel's separation from the Palestinians would likely continue regardless of who would succeed Sharon. Any successor, although perhaps currently hesitant about the proposal, would encounter the same reality that propelled Sharon to abandon his decades-long ideology and embrace the plan for disengagement from the Palestinians. Moreover, any future Israeli prime minister who would share Sharon's assumption that Israel lacks a Palestinian partner for a negotiated end to the dispute would be drawn to self-help measures to preserve Israel's character as a Jewish-democratic state. Equally, such a successor – leading a party that would wish to avoid electoral defeat – would have to be sensitive to the desire of the majority of Israelis to disengage from the Palestinians.

Third, it has become increasingly evident that what was originally presented as a unilateral plan cannot be implemented unilaterally. Indeed, from the outset Sharon sought to coordinate the plan with the Bush administration and to obtain its blessing. Soon after, Egypt was recruited to help mitigate some of the risks associated with the plan, particularly with respect to the security situation in Gaza in the aftermath of Israel's withdrawal. To a lesser extent, British security officials were also engaged to help coordinate the multitude of Palestinian security services. Meanwhile, discussions were conducted with the World Bank in an effort to diminish the likelihood that Israel's disengagement might result in a humanitarian catastrophe, especially in Gaza. Thus, while avoiding direct negotiations with the Palestinian Authority, the Israeli government began a full-scale effort to coordinate the implementation of disengagement with a number of governments and international institutions.

Nonetheless, by mid-2004 a number of critical questions about the proposed separation remained open. From the security standpoint the most important of these concerned the ability to ensure that Israel's withdrawal would not result in complete anarchy in Gaza. In particular, it was not clear whether any Palestinian force – including that commanded by the former Head of Preventive Security, Mohammed Dahlan – would be able to establish a military primacy without Arafat's blessing and support. And, it was also unclear what incentives would likely propel Arafat to grant such support. The precise nature of these notwithstanding, given Sharon's contempt for Arafat, Israel was in any case not likely to cooperate in providing such incentives.

A series of open questions also continued to surround Egypt's role in mitigating security risks. First, it was far from clear how deeply Egypt was prepared to involve itself in Palestinian affairs. More precisely, it was not self-evident that the Mubarak government was prepared to do what it takes to induce Arafat to accept the reorganization of the Palestinian security services and to coerce the Palestinian opposition forces to lay down their arms. Neither was it clear that Israel would be able to contain various inadvertent consequences of Egypt's deep involvement, particularly the danger that Egyptian–Israeli ties would be damaged by continued Israeli–Palestinian friction.

And yet the most difficult questions continued to surround Sharon's ability to overcome the domestic obstacles to the implementation of his already drastically reduced plan. The minority government headed by Sharon in June 2004 was threatened by a further reduction of support should the National Religious Party's remaining government members resign once the implementation actually began. At the same time, Sharon seemed unable to compensate for such expected defections by bringing Labor into the coalition since this could not be done except at the expense of key Likud ministers – Foreign Minister Silvan Shalom, Finance Minister Binyamin Netanyahu, and Education Minister Limor Livnat. Within Likud, objections to joining forces with Labor seemed even stronger than the opposition to the Sharon plan.

Yet these problems pale next to the fear that the evacuation of Israeli settlements in Gaza – and certainly in the West Bank – would be met by physical resistance which may get out of control. In particular, it was not clear that the attempts by the leadership of the settler community to keep the acts of resistance within acceptable boundaries of civil disobedience would succeed. This is because it was far from certain that the extremist elements within the settler community would feel bound by the formal leadership's directives, especially given the split in the spiritual leadership of the community, with some advocating strong physical resistance to evacuation and calling upon IDF soldiers to refuse to obey the expected orders to evacuate.

By the end of June the challenges were compounded by the uncertainty regarding Sharon's coalition. While the formal leadership of the settler community continued to insist that resistance to evacuation must be bound by the rules of democracy, this was prefaced by conditioning such restraint on the willingness to view the decision to remove the settlements as legitimate. Yet with this decision having been rejected earlier by a clear majority among the members of the ruling Likud party, and with the Israeli government making the decision to evacuate resting on a minority among Knesset members, it was far from certain that the settler community would regard the order to "relocate" as legitimate.

While the expected clash was not likely to embroil Israel in a civil war, the possibility that resistance to withdrawal might result in serious casualties among settlers and law enforcement officers alike could not be ruled out. This, in turn, raised the very distinct possibility that Israel would become the scene of domestic traumas, reminiscent of those experienced with the assassination of Prime Minister Yitzhak Rabin in late 1995.

Thus, by mid-2004 Israel's political leadership seems to have experienced the most perplexing paradox resulting from the country's robust capabilities on one hand, and its inability to translate this primacy into solving its dilemmas in the Palestinian sphere on the other. While Israel's impressive gains in almost every dimension of national power and its resulting impressive standing in the region allowed it unprecedented freedom of action, it has become increasingly clear that such strategic supremacy does not provide the Jewish state the capacity to compel the Palestinians to accept Israel's terms for ending the conflict.

With the Palestinian leadership completely de-legitimized and thus rejected as a credible partner, the option of resolving the conflict with the Palestinians through negotiations was dismissed. This, in turn, led to the Israeli government's conviction that until a credible Palestinian leadership would emerge, it would not be possible to ensure Israel's character as a Jewish and democratic state through the negotiated creation of a Palestinian state. This realization led Israeli Prime Minister Sharon to the two significant unilateral measures of building a security fence and beginning Israel's unilateral separation from the Palestinians by disengagement in the Gaza Strip and a northern part of the West Bank. While the construction of the fence continued despite the many problems involved in its precise demarcation, the Sharon's separation plan encountered serious obstacles. The plan's approval by the Israeli government on June 6, 2004 was only a partial remedy, since the governing coalition lost some members as a result, leaving it to rest on a minority among Knesset members. This, in turn, left considerable doubts whether it would be able to overcome the fierce resistance to the proposed withdrawal.

In this sense, Israel has truly reached a strategic crossroads. By mid-2004 its strategic primacy and robust standing in the region allowed it the room for maneuver required to address the dilemmas posed by its Palestinian neighbors through self-help measures. But internally, the fragility of its domestic politics meant that implementation of these measures might expose it to ominous instability. Israel's future as a Jewish and democratic state now seemed to depend on its leadership's ability to weather the expected storms.

CHAPTER ➤➤➤

8

The Intifada: The Dynamics of an Ongoing Crisis

ANAT KURZ

In the course of the current Israeli–Palestinian confrontation, violence once again became a primary Palestinian means for pressuring the Israeli government, as well as an arena for competition over prestige and predominance in the politically fragmented territories. This development was accelerated by the Israeli response to the violence, a systematic military effort that combined punishment and prevention, and was also designed to deter belligerent elements and separate them from the population at large. Palestinian terrorist strikes and the Israeli responses to them occurred in waves; every related outburst could be interpreted as either a response or as a stimulus that would be followed by a response from the other side. In parallel but without any direct dialogue between them, the two sides concentrated increasingly on dealing with the respective implications of the continuing crisis, usually in coordination with foreign elements.

This chapter describes the dynamics that reduced Israeli–Palestinian relations to cyclical violence. The analysis focuses on the dynamics behind the violence; the disintegration of centralized control in the territories, which led to increased pressure for reform in the Palestinian Authority (PA); the failure of the various Palestinian factions to reach an understanding on the goals and methods of the struggle; and the heightened contention between them, particularly in light of the challenge presented by Israel's proposed disengagement.

➤ THE DYNAMICS OF VIOLENCE

When violence erupted in the territories in late September 2000, the PA directed its efforts to managing the events, with the aim of highlighting its historic role in the national struggle. This goal was no less important, and perhaps more so, than the widely perceived objective of the uprising – drawing international attention to the conflict by means of a direct confrontation with the Israel Defense Forces (IDF) and forcing a response from Israel. In an attempt to strengthen the popular image of the uprising, the PA leadership encouraged belligerent elements to escalate the fight. In the early days of the intifada, the role of the various armed bodies – the civilian police, the preventative security agencies in the Gaza Strip and the West Bank, and the factions of the Fatah-Tanzim – evolved from preserving public order to waging the violence. At the outset of the uprising, this policy limited the scope of spontaneous, unauthorized demonstrations and subsequent confrontations with the Israeli security forces. Yet what at the time appeared to be evidence that the PA and its various agencies were in command of the violence, in the long term proved to be a sign of a developing loss of control. Organized terrorist activity not subject to PA central command escalated sharply in the ensuing months. Suicide terrorist attacks were carried out by the organizations of the Islamic movement, the Popular Front for the Liberation of Palestine, and also the al-Aqsa Martyrs Brigades, which grew out of Tanzim branches. By 2001, these attacks, in their showcase and devastating tactics, were spearheading the struggle.[1]

The involvement of the PA in the outbreak of the uprising and in the heightened violence against Israeli civilians and soldiers, both in the territories and within the Green Line, prompted an Israeli effort to puncture the basis of PA control. In addition to the war against the terrorist infrastructures of Hamas, Islamic Jihad, and other opposition organizations, the Israeli security forces concentrated on undermining the preventive security agencies in the West Bank and the Gaza Strip. The result was their accelerated decline, which compounded the freedom of action gained by local factions. The Israeli response to the Palestinian violence escalated further after September 11, 2001. The US administration's subsequent declaration of an uncompromising global war on terrorism was interpreted by the Sharon government as a green light for Israel to destabilize the bases of power of the PA and its leader. Thus the PA became a central and frequent target of IDF attacks, and security agencies headquarters and police stations were among the institutions and symbols targeted. The siege that was imposed on Chairman Yasir Arafat's headquarters was part of the comprehensive campaign against the PA and reflected the effort to dismiss Arafat personally as irrelevant to the Palestinian and international arenas.

Arrests and raids staged in response to terrorist attacks were aimed at disrupting the activity of belligerent elements. Though targeting identified terror activists, attacks on leading figures not infrequently caused casualties and destruction in the surrounding area. Over the course of the confrontation, Israel also regained military control over areas that prior to the uprising had been under PA control. All of these measures and developments disrupted life in the territories, and while both the Israeli and Palestinian economies were affected by the violence, the Palestinian economy, which was significantly weaker at the start, was particularly harmed.[2] Regular transit of workers and goods from the territories was interrupted by the frequent and prolonged Israeli–imposed closures and arduous inspections at roadblocks. Civilian infrastructures were so severely damaged that restoration to their pre-intifada level loomed far in the future.

The violence continued in waves, and the number of casualties soared. As of early March 2004, 931 Israelis[3] and 2,385 Palestinians[4] were killed. The support among the Israeli public for an uncompromising policy against terrorism provided public backing for the IDF activity in the territories and the suspension of security and diplomatic contacts with the PA. The same sentiment was one of the principal factors behind the victories of the Likud Party, headed by Ariel Sharon, in the February 2001 and January 2003 elections.[5] Among the Palestinians, support for the violent struggle in general and for suicide attacks in particular reflected the belief that the continued confrontation would weaken Israel from within, while Israel's militant response to the violence would undermine its international standing.[6]

➤ Undermining the PA's Stature

The intifada returned the Israeli–Palestinian conflict to the international agenda. Israel became an object of severe international criticism due to the scope of its military response to the terrorist attacks, which resulted in numerous casualties and massive devastation of civilian infrastructures. At the same time, no concerted international pressure emerged capable of forcing Israel to withdraw from the territories, and despite the expectations of Palestinian elements, no proposal to dispatch foreign forces to the scene of the conflict was adopted. The Israeli government persisted in its policy of refusing to negotiate under fire, and made an absolute end of terrorist attacks a prerequisite for easing the hardships of the Palestinian population and for the renewal of political talks. However, the decline in the internal and international legitimacy that the PA enjoyed before the outbreak of the confrontation left it incapable of meeting this condition, even if it were willing to.

Notwithstanding the international criticism of Israeli policy, the PA's international prestige was eroded by widespread recognition that its influence in the territories was deteriorating rapidly – largely due to the resumption by Israel of control of the territories. In addition, it was clear that the PA avoided any serious attempt to prevent the terrorism incidents that inevitably drew a sharp response from Israel. Spates of terrorist attacks by the opposition organizations and units of the al-Aqsa Martyrs Brigades, organizationally affiliated with Fatah, further demonstrated the undermining of PA authority and active control. The al-Aqsa Martyrs Brigades formed in the northern West Bank and southern Gaza Strip in a loose network of units. Since 2002, the involvement of Hizbollah and Iran in Brigades activity has grown, while the Brigades' links with the PA security agencies have weakened. The strengthening of the factions not subject to the PA's authority has highlighted the rift that first emerged in the 1980s between the generation of activists born in the territories and the older generation of leaders who came from overseas.[7]

➤ Pressure from Within

In the first year of the intifada, domestic criticism of the PA, audible since the organization's creation, remained peripheral to Palestinian public discourse. In light of the cost of the confrontation in human life and economic decline, however, the corruption and faulty administrative norms resurfaced in public discussion and joined the criticism of the PA's failure to translate the collective effort into diplomatic achievements.[8] Despite the continued support for the struggle in principle, as indicated by public opinion surveys conducted in the territories, the confidence in the PA's ability to achieve gains that would justify the immediate burden and long-term damages of the struggle weakened. Critics demanded a revised strategy that would factor in the strain on the public incurred by the violent methods of the uprising. They also petitioned for the formulation of achievable goals for the struggle, in view of the growing difficulties in the confrontation arena. In the summer of 2002, calls were heard in the territories for an end to the armed conflict with an emphasis on non-violent struggle, along the lines of the early stages of the first intifada of the late 1980s. In June, a group of Palestinian politicians and intellectuals issued a call for reform that included a condemnation of suicide attacks against civilians inside the Green Line. Local popular committees were formed at this time in various locations in the West Bank, mostly in the Ramallah area, according to the model established at the time of the first intifada. These committees attempted to fill the vacuum created by the collapse of the PA civilian infrastructure, and to halt the expanding control of welfare institutions by Hamas.

To a certain extent this internal criticism matched the external demands for reform in government, voiced particularly by the Israeli government and the US administration. Both conditioned renewal of the diplomatic process on administrative and security apparatus changes in the PA. In contrast, however, Palestinian spokespersons stressed the diplomatic stalemate and the harsh Israeli actions as the critical obstacles to reform and democracy in the Palestinian political system. Yet in the summer of 2002, the PA leadership appeared to recognize the need to undertake some administrative restructuring, in order to facilitate the flow of aid into the territories, create jobs, and encourage Israel to take humanitarian action to relieve the distress. The Palestinian leadership also hoped that reform would moderate internal criticism of the PA, rehabilitate its international image, and turn international sympathy and support into pressure on Israel.

➤ External Pressure

During the first year of the intifada, international parties still considered the PA a potential partner for dialogue. Efforts to renew security and political coordination between the PA and Israel began as soon as the intifada broke out, but the election of a Likud government in February 2001 effectively ended the direct talks conducted by the government of Ehud Barak that had attempted to restore the pre-intifada status quo.

In late April 2001 former US senator George Mitchell, who had been appointed by President Clinton as head of the fact-finding committee upon which Barak, the PA, and Clinton agreed in the October 2000 Sharm el-Sheikh Summit, submitted his report. The report, which was prepared in coordination with the UN Secretary General, recommended the renewal of mutual trust and the resumption of negotiations. In June 2001 CIA director George Tenet was sent to the region in order to advance implementation of Mitchell's recommendations. The plan Tenet formulated was designed to lead to a ceasefire, the renewal of security coordination, the return of the IDF to its pre-intifada positions, and resumption of diplomatic dialogue. Israeli and PA defense officials, who met jointly with Tenet, accepted the plan in principle. At this stage, the Israeli government conditioned a return to the negotiating table on seven days of quiet.

The radical Islamic organizations and the independent factions, however, worked to forestall a fulfillment of this condition, and the PA security forces did not make any real effort to stop them.

The US administration, which wished to facilitate the participation of Arab governments in the international coalition against terrorism, or at least avoid their concerted opposition to US efforts in the war on terror, appointed General Anthony Zinni in December 2001 to mediate a ceasefire according to the Tenet plan. Zinni ended a visit to the region in January 2002 without having achieved any results after the *Karine-A*, a ship loaded with weapons and ammunition that the PA had purchased from Iran, was seized in the Red Sea. He returned to the region in March in the wake of escalating violence – in the form of terrorist attacks and IDF operations in the territories – after Sharon withdrew his demand for seven days of quiet. Zinni again failed in his mission, however, because Israel refused to withdraw the IDF from West Bank cities, while for their part, the Palestinians refused to guarantee quiet until the IDF returned to its pre-intifada positions. A suicide attack on March 27, 2002, the first night of Passover, put an end to Zinni's attempts to achieve a formula for mitigating the violence. Two days later, the IDF launched Operation Defensive Shield in the West Bank, an extensive operation that accelerated the ongoing military campaign of deep penetration raids in the territories that had begun several weeks earlier. The operation lasted four weeks and ended with the IDF in control of the cities and refugee camps in the West Bank.

In late March, immediately after the Passover attack, the Arab League summit that had convened in Beirut issued a communiqué outlining principles for an Israeli–Arab settlement, based on a proposal by the Saudi Arabian crown prince, Abdullah. The announcement called for an Israeli withdrawal to the pre-1967 ceasefire lines in exchange for normalization of relations between Israel and Arab countries. As a result of pressure from Syria and Lebanon, it also mentioned the refugee question, saying that the Arab League would accept any settlement on the matter agreed upon by Israel and the Palestinians. The Israelis and the Palestinians were fundamentally divided regarding the details of the plan and the best way to implement it. Even more critical, however, was the gap between the long-range goal of the initiative and the events taking place in the conflict arena. As a result, the proposal failed to gather diplomatic momentum and remained merely a conceptual framework, which the parties could discuss if and when the circumstances changed.

The renewed military conquest of the West Bank accelerated the decline in Israel's international prestige. Nevertheless, the criticism of Israel and sympathy for the Palestinian cause failed to strengthen the PA's standing as a legitimate partner for negotiations. A speech in June 2002 by President George W. Bush, in the context of setting the Middle East stage in advance of the war in Iraq, established a Palestinian state as a diplomatic goal, but insisted that a change of regime in the territories and the election of a leadership not implicated in terrorism was a prerequisite for progress toward realizing this vision. This approach also constituted the basis for the roadmap for peace in the Middle East. The roadmap was initiated by the European Union (EU), which was decidedly less critical of the PA and its leader than the US administration. Yet in view of the emphasis placed by the US administration on counterterrorism, the Quartet, a forum consisting of the EU, the US, the UN, and Russia formed to promote an Israeli–Palestinian settlement, stressed the importance of security reform in the PA as a precondition for resumption of a diplomatic process.

The roadmap outlined three stages to the establishment of an independent Palestinian state in the framework of a general settlement: (1) a halt in violence, and reforms in the PA; (2) general elections and the establishment of a Palestinian state with temporary borders; (3) completion of a permanent settlement in 2005. The Israeli government accepted the roadmap despite its objections to some of the contents, including the emphasis on IDF operations in the territories and continued construction in the settlements as incitement for violence. Since the roadmap was performance-based, however, Israel regarded it as a means to neutralize the rigid timetable for applying the stages toward a settlement. In one key respect, the roadmap suited Israel's position: the insistence that with its current rubric and leadership the PA was not a partner for dialogue, and hence the imperative to remove Arafat from a position of influence through elections in the territories. The foreign ministers of the Quartet approved the roadmap in July 2002, it was endorsed by the US administration the following October, and its implementation was scheduled to begin after the war in Iraq.

Toward Reform in the PA

The combination of internal and outside pressures caused a shakeup in the Palestinian political system. In June 2002, even before President Bush's speech, Arafat presented a 100-day plan, which included the appointment of a cabinet comprised primarily of his followers, joined by a few new professional ministers. On the insistence of the Israeli government and the Quartet, Arafat reorganized the PA security agencies, although he was careful to ensure that the changes would not affect his traditional power bases. In a measure designed to forestall any threat from within to Arafat's status, the PA regional heads of preventive security, Mohammed Dahlan and Jibril Rajoub, were transferred from their positions. This step also testified to the lack of any intention to move against the armed opposition organizations, as demanded by some internal critics. In other words, Arafat avoided making any commitment that he was unable or unwilling to fulfill, and refrained from any meaningful action that could improve the PA's international standing yet would threaten his personal status and that of the PA in the territories. Regarding the economy, on the other hand, real change took place. Salam Fayyad, an internationally recognized economist who had won the trust of the Israeli government and the US administration, was made responsible for economic reform, and in this position, gained control over some of the PA's revenues. Israel welcomed the economic reform in the PA, and released some of the Palestinian tax revenues that had been frozen when the intifada began.[9]

In July, talks took place between then Israeli Minister of Defense Binyamin Ben-Eliezer and Dahlan, Arafat's security adviser at the time, in order to formulate an agreement leading to Israel's withdrawal first from Gaza, and then from Bethlehem. Once more, terrorist attacks and Israeli retaliatory strikes, mostly against targets belonging to the Islamic organizations, rendered the talks void. In September, the cabinet resigned, fearing that it would not win a vote of confidence in the Palestinian Legislative Council (PLC): the security reforms were regarded as surrender to Israel and American pressure, and at the same time, the changes in the cabinet were far from meeting the expectations of internal critics. After a month, however, the PLC approved

a cabinet similar in makeup to the one that had resigned, which as such posed no immediate threat to Arafat's official standing.

Another step toward structural reform in the PA was taken in March 2003, when
the PLC legislated the appointment of a prime minister. Mahmoud Abbas (Abu
Mazen) was appointed to form a cabinet. A prominent member of the older generation of the PLO, Abu Mazen was an outspoken critic of the violent intifada, yet as one
of the PA personalities from outside the territories, he lacked any independent power
base. At the time of his appointment and while a new government was in formation
just before the war in Iraq, terrorist activity declined substantially. Israeli security
sources assessed that renewed Israeli military control of the territories, the persistent
hunt for Hamas members and other belligerent elements, and the undermining of the
PA's centers of power had achieved their goal. In order to recreate PA grass roots
control and protect the drop in violence, Abu Mazen sought to appoint Dahlan as
minister of internal affairs, responsible for security. Arafat, however, only approved
Dahlan's appointment to head the PA security apparatus. The national security forces
– the Palestinian army which was transplanted from Tunisia upon Arafat's return to
the territories – as well as the general intelligence, regional governors, and Tanzim
remained under Arafat's control, out of reach of the minister of internal affairs. In May
2003 with the help of the Quartet, which pressured Arafat to forego the right to appoint
and fire ministers at will, the PLC approved the new government, heavily populated
by Arafat's longstanding loyalists. Abu Mazen's most significant test, however, still
lay before him.

➤ THE POLITICS OF VIOLENCE: FOR AND AGAINST A *HUDNA*

Negotiations between the Palestinian organizations to reach agreement on the nature
and goals of the struggle, as well as the failure of these contacts, were among the key
indications of a weakened PA. These talks, which attempted to forge understandings
between the factions, took place both inside and outside of the territories. The talks
were accompanied from an early stage by external pressure and mediation, underlying
the difficulty of achieving a direct dialogue between the various parties. The specific
goals of the talks evolved according to developments in the conflict arena, the external
environment, and the internal Palestinian scene. In any case, Israel was not a party to
these discussions but an interested spectator who influenced their development and
whose policy was affected by their results.

The questions raised at the beginning of the talks that consistently remained open
were whether the belligerent organizations would refrain from terrorist attacks that
provoked Israel's standard responses to violence, and whether the creation of any
effective civilian and military authority would be possible in the territories from which
Israel withdrew. There were in effect three sets of talks, distinguished by the shifting
balances between internal Palestinian and international exigencies. The first series of
talks was designed to foster unity in the struggle and stem the PA's loss of control in
the confrontation theater. For the institutional leadership in the territories, the
purpose of the talks was to reinforce its position at home, while the opposition factions
joined the talks in order to safeguard their ability to veto any compromise with Israel.
The second series was intended to reach an agreement that would make it possible to

restrain the violence and thus enable a renewal of Israeli–Palestinian dialogue toward a resolution, which depended on a pause in the violence. As far as the PA was concerned, the understanding between the various factions was intended first and foremost to strengthen its international prestige and in so doing rehabilitate its domestic authority. The third set of talks resembled the first in that it was primarily aimed at reviving the PA's standing on its home ground. This time, however, the talks took place against a background of the collapse of central authority in the territories and the escalating struggle for political primacy among the factions.

➤ Unity of the Ranks

The first set of talks, which opened in early 2001 and continued intermittently for approximately one year, aimed to unify the factions in light of the developing reality in the conflict arena. From a Palestinian perspective, this was dominated by the distress of the population in the territories and the weakening of the PA's control during the uprising. The escalation of the violence outside the control of the PA and fear that the intensification of the Israeli–Palestinian conflict would instigate upheaval in their own territory led the Egyptian government, together with Saudi Arabia and Jordan, to work toward formulating understandings between the PA, Fatah, Tanzim, Hamas, Islamic Jihad, and the other opposition factions on the methods to be used in the conflict. In mid-2002, the EU became increasingly involved in the attempts to formalize the talks between the various factions.

For the PA, which was represented in the talks by Abu Mazen, then Fatah general secretary, a ceasefire was a means of stabilizing its position at home. The PA therefore tried to reach understandings that would on the one hand encourage the Israeli government to withdraw the IDF to its pre-intifada positions, and on the other hand concomitantly halt the economic decline in the territories, strengthen the Palestinian public's endurance, and in turn soften criticism of the overall PA performance. Hamas and the other Palestinian opposition organizations participated in the talks for lack of choice, in order to prevent the PA from acting against them and devising a policy they opposed, although the renewal of diplomatic dialogue with Israel was only a secondary goal for the PA. Indeed, the very participation of Hamas in the talks constituted recognition of its internal Palestinian standing, in addition to its proven ability to use terrorist attacks to affect the course of the conflict.

The meetings that began in Cairo in November 2002 marked a new phase in the talks, featuring more intensive Egyptian involvement in intra-Palestinian relations as well as the definite positioning of the Fatah-affiliated al-Aqsa Martyrs Brigades in the ranks of the opposition, together with Hamas and Islamic Jihad. At the top of the agenda lay a possible *hudna* – an agreement between the organizations on a temporary halt in the violence for a specific period of time. With the war in Iraq looming in the imminent future, these talks reflected the concern that the erosion of the PA's international standing would render impossible any deflection of the pressures expected from the US administration after the war. The Quartet and the Saudi Arabian government supported the Egyptian mediation effort. The initiative also won the consent of Syria, which wished to be removed from the US State Department list of states sponsoring terrorism. For that reason, Khaled Mashal, head of the political wing of Hamas who resided in Damascus, was permitted to participate in the talks.

Significantly, Hamas joined the talks at a time of its own increasing economic hardship. Measures adopted by the United States and Britain after September 11, 2001 substantially curtailed the transfer of donations by charity associations to the territories.[10] The organization's leadership compensated for the missing funds by tightening their relationship with Tehran, while trying to ease Israeli pressure on the Hamas civilian infrastructure: calls were concurrently issued in the Gaza Strip for Hamas to cease firing rockets at Israel, in order to deny Israel the pretext for raids and destruction of infrastructures.[11]

The declaration by Sharon during the election campaign in Israel that a permanent settlement between Israel and the Palestinians would include the establishment of an independent Palestinian state heightened the interest of the Hamas leadership in preventing the renewal of political dialogue. Yet disputes within Fatah itself concerning terms for agreement on a *hudna*, as well as the Israeli air force attacks against Hamas' military infrastructure, interrupted the inter-organizational dialogue. Under Egyptian pressure, however, a "National Covenant" was drafted in February 2003. The agreement provided for coordinated action to end the occupation, dismantle Jewish settlements, found a Palestinian state in territories released by Israel with a capital in Jerusalem, defend the right of the refugees' return, and fulfill UN decisions in a manner that would guarantee a just and lasting peace between Israel and the Palestinian people. The issue of disarming the belligerent organizations – an unequivocal Israeli prerequisite for tacit backing of understandings between the Palestinian organizations – was not even mentioned.

This covenant ended up having no practical meaning. Already by late January 2003, before it was published and after a period of relative quiet – which strengthened assessments in Israel that the peak of violence had passed – a suicide bombing on a Jerusalem bus by the al-Aqsa Martyrs Brigades highlighted the PA's loss of control over factions affiliated with Fatah and Tanzim. For its part, Hamas rejected pressures to restrain its military wing, Izz a-Din al Kassam, and accede to the PA's demand that attacks be confined to the West Bank and Gaza Strip. In addition to the persistent lack of agreement on the borders and character of the future Palestinian state, representation of the Islamic movement in the PA's institutions remained a bone of contention between the organizations.

➤ The Ceasefire

The ceasefire declared in late June 2003 followed a series of further contacts between the PA and the opposition organizations, which peaked in a few weeks of intensive talks held under intense international pressure. At the beginning of the month, Abu Mazen, Ariel Sharon, George W. Bush, and Jordan's King Abdullah held a summit in Aqaba to mark the beginning of the implementation of the roadmap. The speech by Abu Mazen included a commitment to halt the violent uprising and disarm the opposition's organizations. From Aqaba, Bush, Abdullah, and Abu Mazen traveled to Sharm el-Sheikh, where they met with Egyptian president Husni Mubarak, the Saudi Arabian crown prince, and King Hamad bin Isa al-Khalifa of Bahrain, in order to ensure regional support for a ceasefire that would launch the first stage of the roadmap. The PA and Hamas, followed by Fatah, the al-Aqsa Martyrs Brigades, Islamic Jihad, and the Democratic Front for the Liberation of Palestine – separately, since no joint announcement was published – declared their willingness to suspend terrorist attacks against Israeli soldiers and civilians for three months, both in the territories and within the Green Line. Fatah, for its part, undertook to halt terrorist attacks for six months.

For Abu Mazen's government, a pause in terrorist attacks was an existential necessity.[12] A halt to the violence could strengthen the PA's internal standing by improving its international prestige, especially if the Quartet regarded the lull as the PA's successful fulfillment of its obligation to restrain violence, and on that basis demand that Israel carry out its part of the first stage of the roadmap. In exchange for a ceasefire, the PA expected Israel to release prisoners, relax closures and sieges, withdraw from urban centers, cease deep penetration raids, halt the pursuit of wanted terrorists, freeze settlement construction, and evacuate illegal outposts. The respite was also expected to secure the release of funds frozen in Israel and generate a flow of European aid, thereby facilitating the rehabilitation of both government institutions and civilian infrastructures in the territories. In addition, the ceasefire, based on an agreement among the organizations was designed to: demonstrate, if only for appearances' sake, unity in the PA leadership; allow the rebuilding of the Palestinian security agencies; and reestablish the PA's monopoly over the use of weapons. Above all, the ceasefire was expected to help the PA dispel the Israeli claim that quiet, unaccompanied by the disarming of belligerent opposition groups, did not meet the basic conditions for beginning implementation of the roadmap.

The possibility that the pause would allow the Abu Mazen government to establish itself on a firm basis was the price that Hamas and the other opposition organizations had to pay in order to survive. The resumption of the diplomatic channel, particularly in light of Abu Mazen's speech at Aqaba, was considered a threat that necessitated cooperation with the PA. The escalating IDF pursuit of active members of the terrorist front, including the attempt to kill Abdul Aziz Rantisi, Sheikh Yassin's deputy, also played a role in convincing Hamas to declare its willingness to suspend terrorist attacks.

At the end of the talks between Dahlan and then IDF Coordinator of Government Activities in the Territories Major General Amos Gilad, which were driven by American and Egyptian pressure to embark on the roadmap, principles of an Israeli–Palestinian agreement were hammered out. The agreement provided for the transfer of responsibility for security in the Gaza Strip and Bethlehem to the Palestinians, and Israel's suspension of military activity in the territories. In exchange, the PA was committed to act against terrorism. Officially, Hamas and Islamic Jihad were not obligated to the understandings between the PA and Israel, just as the PA was not a formal party to the decision by Hamas and Islamic Jihad to suspend their violent struggle. The agreement reached by Israel and the PA, however, could not have been concluded without understandings between the PA and Hamas. These were reached when Abu Mazen and Dahlan promised the Islamic organizations that they would not be disarmed, that the ceasefire would be conditional on a release of prisoners from Israeli jails, and that Abu Mazen and Dahlan would try to end the targeted killings by the IDF. Significantly, these two Palestinian leaders were driven by considerations that had previously prevented and would later prevent again a head-on confrontation with the belligerent opposition, headed by Hamas. This balance, whereby steps designed to enhance international legitimacy were likely to undermine domestic legitimacy, eliminated any possibility of forcibly restraining the belligerent elements and renewing security coordination with Israel, let alone resuming the diplomatic process.

The ceasefire came into effect at the beginning of July and was followed by a few weeks of relative quiet. Attempted attacks along the Green Line occurred less frequently. Izz a-Din al Kassam cells, spurred on by the Hamas leadership in Damascus, continued to operate, but the local leadership was more concerned about reconstructing its damaged social and military infrastructure. Dahlan ordered the closing of the tunnels in the Rafah area used to smuggle weapons from Egypt, but this move was not followed by a decline in the smuggling, let alone a comprehensive campaign to curb violence. Among other reasons, this was because the national security forces blocked systematic operations to stop terrorist strikes and rocket attacks into Israeli territory from the northern Gaza Strip, which had been transferred to Palestinian security control. The al-Aqsa Martyrs Brigades proved that restraining them was no easier than forcing Hamas to observe the ceasefire. In order to encourage the independent factions to suspend terrorist attacks and to sever the tightening relations between them and Hizbollah and Iran, Dahlan acted to co-opt their activists into the PA agencies. Money was distributed as a reward for not carrying out attacks. The large number of armed groups, however, made centralized control impossible. The overlapping of command and the lack of coordination between the various security agencies prevented law enforcement, which, over time and subject to a renewal of the diplomatic process, might have circumscribed the violence.

Within a few weeks after the *hudna* was declared, incitement and support for continuing the confrontation declined in the Palestinian street. The transit of people and goods through roadblocks was eased, licenses to work in Israel were increased, more than 300 prisoners were released from Israeli jails, and targeted killings were suspended.[13] Israel, however, was not satisfied with Dahlan's measures, and demanded concrete steps to disarm the belligerent factions. IDF activity in the territories – raids and arrests – continued. More than once, operational tactics dominated overall considerations, while the residual effects on the general cycle of violence were largely ignored. Humanitarian gestures by Israel, which could have enhanced Abu Mazen's popular image as a champion of the Palestinians, were of rather limited scope. Illegal outposts were not removed, as promised by the Israeli government's commitment to the roadmap, and overall Abu Mazen was prevented from demonstrating his effectiveness in promoting an end to the occupation.

The attempted assassination of Rantisi on August 20, 2003 came one day after a terrorist attack on an IDF post at the Erez border crossing and one day before a suicide terrorist attack on a bus in Jerusalem, carried out jointly by Hamas and Islamic Jihad. These events signaled the end of the ceasefire. The Israeli government decided to target Hamas leaders, leaving none of them immune. At the same time, it continued to demand the disarming of the Palestinian organizations, a move that could not be accomplished without a frontal confrontation among Palestinians. Hamas and Islamic Jihad renounced the *hudna*. Arafat, Abu Mazen, and PA Minister of Planning and International Cooperation Dr. Nabil Shaath called for its renewal, but within a few weeks, it was evident that the inter-organizational dynamic had not changed. In a measure designed to demonstrate his control at home, Arafat ordered the confiscation of twelve of the funds that financed Hamas welfare activities and invited the needy to apply to PA offices. In view of the PA's new apparent determination to restrain terrorism and the resumption of Israel's policy of targeted killings, Hamas leaders went underground and tried to renew discussions of a *hudna*. Other factors motivating them included the fear that they would lose popular support, given the hardships of daily life in the territories; the worsening of Hamas' economic situation caused by its classification as a terrorist organization by the EU; and the suspension of aid operations by the Arab world.

Abu Mazen resigned in mid-September. He blamed the collapse of his government to a large degree on Arafat, who prevented the consolidation of the Palestinian security forces and did not promote a ceasefire, which would have obliged Israel to suspend IDF operations in the territories. The failure to formulate an enforceable position regarding a ceasefire was evidence of the fragmentation in the domestic Palestinian scene. The limited responsiveness of the Israeli government to the attempts to restrain terrorist attacks under these circumstances and the irrelevance of the roadmap to events in the conflict arena rendered the Palestinian theater all the more volatile.

➤ A Question of Control

Following Abu Mazen's resignation, formation of a new government was entrusted to Ahmed Qurei (Abu Ala), spokesman of the Palestinian Legislative Council. Abu Ala, for his part, left decisions about appointments to Fatah institutions – a signal of his intention to avoid any confrontation with Arafat and the PLO's older generation. In early October 2003, following another wave of violence in the territories and after the Israeli cabinet decided in principle to expel him, Arafat ordered the formation of an emergency government.[14] Bearing arms in the streets of the Gaza Strip was officially forbidden, although the PA security forces did little to enforce this order.

The emergency government was sworn in on October 7, 2003. Two days earlier, the Israeli air force attacked an Islamic Jihad base near Damascus, in response to a suicide terrorist attack in Haifa. On November 12 the government of Abu Ala, consisting mostly of Arafat loyalists, was sworn in. Except for maintaining the new position of prime minister, the composition of the government reflected no structural reform in the PA. In this case also, no overall control of the various security forces was given to the prime minister or the minister of the interior. The incoming prime minister announced that he intended to implement the roadmap and try to promote a ceasefire among the organizations. In contrast to his predecessor, however, Abu Ala refrained from making a commitment to halt the violent intifada. Aware of the most significant challenge facing him, he promised to gain control of the independent armed elements. Nonetheless, this task, which in the early stages of the intifada was believed to be less difficult than restoring Israeli–Palestinian trust and reconstructing the diplomatic process, proved to be extremely difficult, if not impossible, to accomplish.

Upon taking office, Abu Ala initiated discussions to renew the *hudna*. Islamic Jihad and Hamas initially rejected calls to halt terrorist attacks until the formulation of understandings for an official ceasefire, and contacts were not revived in Cairo until early December 2003. As in previous rounds of talks on a *hudna*, without Egyptian pressure and mediation the various elements could not have been brought together. This time, the Egyptian initiative, supported by Damascus, was formulated under pressure from the European Union and the United States to help establish quiet in the territories.

Even before the December talks, meetings took place between Abu Ala and Israeli defense representatives with the goal of reaching tacit understandings, yet with no illusions of reaching a binding agreement. Indeed, without support for a ceasefire among all Palestinian factions, the meetings laid no foundation for a renewal of security coordination. In return for a halt in terrorist attacks, the PA demanded the removal of roadblocks, a settlement freeze, the release of prisoners, and a suspension of IDF activity in the territories. Hamas' conditions did not differ in principle, but were not presented as a basis for dialogue with Israel. Hamas representatives demanded that the PA release frozen Hamas funds and guarantee the participation of its representatives in elections for local authorities when held in the territories.

Added to this list of familiar demands was a demand for a halt in the construction of the separation fence in the West Bank, approved by the Israeli government in April 2002, and the demolition of what was already erected. The Israeli government depicted the fence as a barrier to thwart terrorists intent on staging attacks, particularly suicide attacks within the Green Line, and therefore essentially a security measure. The demarcation of the fence was approved by the Israeli government in October. Much of the fence was to run east of the Green Line, while certain parts surrounded large settlement blocs, effectively dividing areas in the West Bank and enclosing Palestinian towns with a wall. The fence became a bone of contention between Israel and the United States because it was impossible to separate its declared security objectives from its political image as a unilateral measure for establishing territorial facts on the ground.

The discussions between the Palestinian organizations reached an impasse several days after they began. The immediate pretext for their interruption was the announcement on December 4, 2003 of the Geneva initiative, a product of track-II Israeli–Palestinian thinking. The initiative, which outlined principles for a permanent settlement and set mutual conditions for reining in the cycle of violence, drew criticism from both Israelis and Palestinians, as well as serious interest in its particulars. Fatah disavowed the initiative, and the Islamic organizations, which stressed the contradiction between those understandings and the talks in Cairo, withdrew from the talks as soon as it was published. However, when the assassinations of terrorist leaders resumed after being suspended by Israel during the Cairo talks, the Hamas leadership proposed a compromise that would exempt Israeli civilians from the confrontation for a period of three months. At the same time, Hamas rejected an Egyptian proposal for a general unilateral ceasefire.

Concurrently, the PA's security forces began hesitantly to collect unauthorized arms and to seal tunnels in the Rafah area, but effectively gave up the idea of a direct and sweeping confrontation with the opposition organizations and armed factions. Abu Ala argued that suicide terrorist attacks gave Israel a justification for using collective punishment in the territories and refusing to dismantle outposts, but merely issued after-the-fact condemnations, without publishing a statement of principle on the matter. Since the PA itself could not ensure the suspension of terrorism nor ensure that a ceasefire would not be exploited for the purpose of organizing hostile elements, the PA requested US guarantees for a suspension of IDF actions in the territories.

In addition to the traditional disputes between the Palestinian organizations, another obstacle to a ceasefire lay in the operations of independent elements that had broken free of any institutionalized framework and had either disavowed their organizational loyalties or had emerged autonomously. Demonstrative parades of armed activists shooting in the streets and the violent settling of personal and political accounts became routine events. Entire neighborhoods in cities and refugee camps in the West Bank were controlled by factions affiliated with the al-Aqsa Martyrs Brigades. In the Gaza Strip, Islamic groups became a factor deterring attempts to enforce central authority, particularly measures to collect unauthorized arms.

➤ THE CHALLENGE OF DISENGAGEMENT

The inability to implement even the first stage of the roadmap pushed both Israelis and Palestinians to unilateral, self-serving measures that would reduce the problems of their respective populations created by the years of violence. Israel proceeded with the construction of the separation fence, which it portrayed as an indispensable measure to contain the security situation. On the Palestinian side, restoration of public order and renewal of the PA's authority became the most important goals. As in the past, efforts in this direction were made through external mediation, mostly by Egypt.

Intra-Palestinian power struggles, particularly between the remnants of the security agencies on the one hand, and the independent factions that had seceded from Fatah and members of Hamas and Islamic Jihad on the other, escalated against the backdrop of intensified public discussion in Israel of a unilateral disengagement from the Gaza Strip. This idea, first raised by Ariel Sharon in November 2002, was further evidence that Israel had abandoned the option of an agreed settlement, and was interpreted as an acknowledgment of the limited ability of military control to grapple with terrorism successfully. The political and security debate in Israel focused on the disengagement plan, which was presented as both a security solution and as a political-territorial measure. The plan joined the security fence as a focus of the Israeli–American dialogue. Both the fence and the disengagement plan won significant support in Israeli public opinion, and both aroused bitterness and anxiety in the Palestinian public.[15]

While opposition to the fence remained general and a matter of principle, however, a trend evolved in Palestinian public opinion toward accepting the idea of disengagement. Preparations for the period following the Israeli withdrawal from the Gaza Strip were launched. In mid-March 2004, in coordination with Egyptian and British security officials, the PA presented a multi-stage plan for enforcing order. Concurrently, belligerent elements in the Gaza Strip redoubled their efforts to prove that withdrawal would be conducted under pressure. For their part, Hamas and Islamic Jihad delayed their response to the plan, and Hamas formulated its own proposal. Sheikh Yassin announced that Hamas would consider halting terrorist attacks in the Gaza Strip if the IDF indeed withdrew, but warned that terrorist attacks would escalate in the absence of a corresponding withdrawal from the West Bank. Hamas likewise did not disavow the intention in principle to achieve coordination with the PA. As Hamas had always done since it was founded, however, its leaders again rejected the invitation to join the national leadership, since its demand for representation proportionate to popular support for the Islamic movement was denied. Nor did Hamas try to assert its responsibility for security in the Gaza Strip. Hamas left this task, which would inevitably be interpreted as surrender to Israeli pressure and a move toward dialogue with Israel, to the PA.

In late March Abu Ala declared that disengagement could prove an opportunity to resurrect the diplomatic process. This statement, which was issued in an attempt to calm the raging tempers and growing anarchy in the Gaza Strip, followed a series of events that reflected further that the dynamics of confrontation have become rooted in the conflict.

A suicide terrorist attack in Israel's Ashdod port on March 14, 2004 led to the cancellation of a scheduled meeting between Ariel Sharon and Abu Ala. Each leader had planned to use the meeting in his own way to show that he was not at fault for the diplomatic stalemate. Hamas and the al-Aqsa Martyrs Brigades jointly carried out the terrorist attack, with help from Hizbollah. The attack, which was designed to demonstrate the limitations of the separation fence and enhance efforts to portray an Israeli disengagement as withdrawal under fire, was evidence of the disintegration of Fatah control over its own factions in the Gaza Strip. As in the past, however, Arafat refused to permit the arrest of Fatah personnel in the Strip. The Israeli security cabinet resolved to escalate the attacks against Palestinian leaders implicated in terrorism, in order to deliver a deterring message and to counter accusations by right wing circles in Israel that withdrawal constituted surrender to terrorism. The assassination of Sheikh Yassin on March 22, 2004 illustrated the seriousness of this intention. Calls for revenge were voiced in the territories, and the funeral procession for Yassin turned into a demonstration of power by Hamas. A request by Palestinian public figures, including PA officials, to exercise restraint and abandon the armed struggle was officially rejected by the Islamic organizations and found no support in the territories.

The rift between the Israeli government and the PA and the PA's difficulty in reconciling with the opposition forces expedited international efforts to help the PA overcome anarchy, particularly in anticipation of Israeli disengagement. The mood in the territories following the assassination of Sheikh Yassin undermined the challenges of Arab governments to resume the diplomatic process, and served as one of the excuses for canceling the Arab summit scheduled in Tunisia to discuss the Saudi initiative, among other issues. At the same time, concern over the possible spread of anarchy from the Gaza Strip to Egypt caused the Egyptians to persist in their effort to consolidate the PA security agencies. In this context, measures were taken to defuse tension between individuals in Fatah, particularly Dahlan and Rajoub, whose mutual cooperation was considered essential for an effective counterterrorism campaign. Pressure was applied on Arafat to permit a reorganization of forces. For its part, the Israeli cabinet acted to coordinate the separation fence project with the US. Objections by the US administration to the route of the fence, since it was regarded as liable to complicate the establishment of a viable Palestinian state with reasonable territorial contiguity, led to an amended demarcation. The US adherence to its Middle East vision also explained the administration's demand that disengagement from the Gaza Strip be part of a plan that would include dismantling of settlements in the West Bank.

Debate over Sharon's proposed disengagement plan in Israel, the territories, and around the world – even though the plan was neither endorsed by the Likud Party nor was debated in the Israeli cabinet and the Knesset – reflected the assessment that its implementation was only a matter of time. Discussion therefore centered on preparations for the period following disengagement, rather than on the merits of disengagement itself. In early April 2004, with Egyptian mediation, the Palestinian organizations formulated yet another draft understanding concerning cooperation to achieve order in the areas from which the IDF would withdraw. In contrast to Egypt, which acted to foster Palestinian unity in the belief that law enforcement would be impossible without Hamas, the US administration warned against including Hamas in the Palestinian national leadership, since Hamas was classified as a terrorist organization. This position ignored the fact that curbing terrorism without Palestinian unity would be no less difficult than it had been after the roadmap was first launched, when the PA was unable to stop violence due to a lack of coordination between its forces and Hamas. Since then, the security and legal institutions in the PA reached a state of collapse, and popular support for Hamas increased significantly.

Hamas itself has also experienced a severe leadership crisis. The assassination by the IDF of Rantisi, the heir of Sheikh Yassin, on April 17, 2004 was designed to disrupt preparations for terrorist attacks against Israel, but the decimation of the Hamas leadership in the Gaza Strip strengthened its leadership in Damascus, which has close ties with Iran. This reduced the already slim chances of reaching a binding agreement between Hamas and the PA concerning a ceasefire.

In the territories, restraining the violence was in any case no longer considered part of an agenda shared by the PA, Israel, and the Quartet, but as a Palestinian national goal in its own right. In any case, the efforts to reach an agreement between Palestinian organizations that would restore order in the territories proceeded in parallel to the measures adopted by Israel for dealing with the crisis. The respective efforts reflected the fact that their link with any coordinated diplomatic plan along the lines of the roadmap was becoming increasingly tenuous.

➤ LOOKING FORWARD

As was made clear in the early months of the struggle, the cycle of violence could not be easily broken because Israel and the Palestinians held different conceptions of why the conflict escalated. Israel stipulated a halt in terrorism as a condition for canceling IDF activity in the territories, while the PA insisted on the IDF returning to its positions before the outbreak of violence as a precondition for restraining terrorism. The renewed occupation of the West Bank and the Gaza Strip by the IDF weakened the PA's control there and promoted the increased freedom of action by elements not recognizing the authority of any national institution. The leadership changes in the PA, instituted as a result of internal and external pressure, neither led to a change in the nature of the struggle nor enhanced the PA's ability to impose its authority on belligerent opposition elements. Terrorist strikes by opposition elements stimulated Israeli military reprisals and prevented measures designed to calm the situation, halt the collapse of the civilian infrastructure, and enable a resumption of negotiations. The PA's grip on the territories loosened, while no other party arose in its place that was capable of adopting an alternate strategy for leading the struggle and presenting the Israeli government with a substantial political challenge.

From an Israeli perspective, the prolonged crisis strengthened the perception that emerged already in the first weeks of the crisis that Israel had no negotiating partner, regardless of the particulars of any proposed settlement. At the same time, the extended confrontation drew attention to strategic threats; the diplomatic process, even if bumpy and accompanied by violent events, had previously made it easy to postpone discussion of these issues. The chief threats raised were the demographic consequences of continued occupation and the preservation of the settlement enterprise in its entirety, i.e., Israel's future as a Jewish state, and the social consequences of continued control over millions of Palestinians, i.e., Israel's future as a democratic country. Recognition of Israel's contribution to the cycle of violence, through its ongoing presence and massive military operation in the territories, and especially the limited degree to which Israel's military power could be channeled to repress terrorism significantly, also emerged. This led to a change in the conclusions drawn from the years of confrontation. Palestinian terrorism remained the main reason for the rift between the Israeli government and the PA, but from the perspective of the Israeli public as reflected in opinion polls, the importance of terrorism as a factor in continued military control of the territories and preventing an organized withdrawal lessened.

In the Palestinian territories themselves, the accumulated damage caused by the conflict – the casualties, destruction of infrastructures, and collapse of a central authority – increased support for the violent struggle and the opposition organizations, as well as affiliation with factions that had broken off from a command hierarchy. At the same time, there were calls in the territories to abandon the armed struggle and continue the struggle through non-violent means. Also voiced occasionally was the possibility of altering the balance of forces between Israel and Palestinian society through demographic pressure, that is, by waiting for the day when Palestinians would constitute a majority between the Mediterranean Sea and the Jordan River and win control over the country through elections on the basis of "one man, one vote."

This solution for the crisis, however, born of the realization that the popular struggle could not be converted into diplomatic and territorial achievements despite international sympathy for the Palestinians, did not alter the public agenda. Belligerent elements, which persisted in their effort to stimulate an Israeli military response, dictated the events. International pressure did lead the PA to attempt to achieve an inter-organizational understanding regarding a ceasefire in order to halt IDF operations in the territories and facilitate a renewal of diplomatic talks. In view of the failure to reach a compromise between the organizations and the anarchy prevailing in the territories, however, the renewal of dialogue with Israel became secondary to the restoration of public order.

The understandings reached between the Bush administration and Ariel Sharon regarding Israeli disengagement from the territories, publicly announced in mid-April 2004 and approved by the House of Representatives and Senate two months later, included an agreement on the final goal of establishing a Palestinian state in the Gaza Strip and the West Bank. The main parameters for establishing such a state, however, were decided independently of Palestinian input. These parameters featured precluding the return of refugees to the State of Israel, the acceptance of settlement blocs on the West Bank, and the method of implementing the plan, that is, in accordance with the scope and timing of the Israeli withdrawal.

As could have been predicted, Sharon's plan aroused deep concern in the territories and some sharp criticism in the international arena. In addition, it also elicited angry responses within the Likud Party. Rejection of the plan by the Likud on May 2, 2004 hindered Sharon's intention to present it the Knesset and prepare the political and military grounds for its implementation. However, the political commotion over the plan did not eclipse the momentous change within the Israeli public over the territorial status quo. The metamorphosis that occurred once the Oslo process was replaced by a direct and bloody confrontation was reflected in the public support for Sharon's disengagement plan. For many Israelis, Israeli–Palestinian cooperation in making and implementing decisions was no longer deemed a prerequisite for what was considered a necessary withdrawal from the Gaza Strip and the West Bank.

Notes

1 From the beginning of the intifada until August 2003, there were 18,125 terrorist attacks, of which 9,628 were in areas bordering or inside the Gaza Strip. The fence surrounding the Gaza Strip was cited as the cause of the relatively low number of casualties in the neighboring areas (10 percent). In terms of numbers, the suicide attacks accounted for a small fraction of the total number of terrorist attacks. According to the IDF spokesperson <www.idf.il>, few suicide attacks took place in most months of the intifada – between two and four. The peak period was December 2001–June 2002, and seventeen suicide attacks occurred in March 2002. By early March 2003, however, 56 percent of the Israeli fatalities were killed in 116 separate suicide attacks (Ze'ev Schiff, *Ha'aretz*, August 3, 2003).

2 According to a report published by Palestinian Minister of Finance Dr. Salam Fayyad, economic activity in the territories dropped by 50 percent since the outbreak of the intifada (*Globes*, March 2–3, 2003). According to another estimate, activity fell 31 percent (*Globes*, October 14–15, 2003). Also according to *Globes* (April 22–23, 2003), while international aid to the PA in 2002 totaled approximately $700 million (with merely $500,000 from Arab countries, *Ha'aretz*, September 26, 2003), only 10 percent was designated for infrastructures and the educational and health systems. The remainder was added to the budget for financing the salaries of approximately 125,000 public employees, of whom 56,000 were security personnel. The unemployment rate in the territories tripled from 21 percent in 1999 to 60 percent in 2002 (*Globes*, October 14–15, 2003). One factor thought to have played a key role in the continued economic activity, even under these conditions, was the relative durability of the banking system in the territories. A report published by the Bank of Israel (*Ha'aretz*, March 31, 2004), claimed that the intifada was one of the main reasons for the economic recession in Israel. The damage to the Israeli economy, not including defense spending, reached 2–8 percent of the GDP, amounting to 31–40 billion NIS. See also Imri Tov, "Economy in a Prolonged Conflict: Israel 2000–2003," *Strategic Assessment* 6, no. 1 (2003): 20–25.

3 Figures from the IDF spokesperson, March 11, 2004, http://idf.il/english/news/ jump_2_eng_300900.stm.

4 Figures from B'Tselem (The Israeli Information Center for Human Rights in the Occupied Territories), <http:/www.btselem.org/English/Statistics/Al_Aqsa_Fatalities_Tables.asp.> According to B'Tselem figures, 839 Israeli had been killed as of this date. According to the IDF spokesperson (Ze'ev Schiff, *Ha'aretz*, August 3, 2003), 551 of the Palestinian fatalities were terrorists, and the rest civilians.

5 According to a public opinion survey on national security issues (Asher Arian, *Israeli Public Opinion on National Security 2003*, JCSS Memorandum No. 67, Tel Aviv: Jaffee Center for Strategic Studies, Tel Aviv University, October 2003), approximately 90 percent of the respondents in surveys conducted in 2001, 2002, and 2003 supported the killing of terrorism activists as long as negotiations had not been renewed. Seventy percent supported closures

and economic sanctions in the territories during those years. Fifty-seven percent of the respondents supported re-entering Area A in 2001, a figure that rose to 72 percent in 2002. The trend continued in 2003, with support rising to 76 percent. In 2002, 57 percent of respondents said that the government's policy in the territories was too soft, 30 percent said it was appropriate, and 9 percent said it was too harsh. In 2003, 29 percent said government policy in the territories was too soft, 56 percent said it was appropriate, and 13 percent said it was too harsh. Evidence of support by the Israeli public for a harsh policy was provided by the degree of support for the assassination of Ahmed Yassin, which reached 60 percent, while 81 percent of the respondents in the same survey (*Yediot Ahronot*, March 23, 2004) believed that terrorism would increase as a result of the killing.

6   According to an opinion survey in the West Bank and Gaza Strip by *PSR – Survey Research Unit*, the proportion of support for the conflict, based on the belief that it would achieve political goals, came to 70 percent in July 2001. This figure is particularly significant in view of the general recognition – 93 percent – that the intifada had harmed the Palestinian economy. Sixty-one percent of the respondents in a survey published in December 2001 were of the opinion that the violent confrontation would promote Palestinian national goals more than negotiations. A survey published in August 2002 showed 70 percent support for this view. In a July 2001 survey, 63 percent of the respondents supported an immediate return to negotiations, although 46 percent believed that the peace process was dead, 92 percent supported the confrontation with the IDF in the West Bank and Gaza Strip, and 58 percent supported terrorist attacks against Israelis within the Green Line. The survey published in December 2001 reflected similar trends, with slight differences, while the survey published in May 2002 indicated the emergence of a change in the attitude toward the PA and the nature of the struggle against Israel. Support for terrorist attacks inside Israel fell to 52 percent, although 86 percent of the respondents opposed the arrest of those perpetrating attacks within Israel. According to a *PSR* survey published by the *Jerusalem Post* on March 28, 2004, 87 percent of the respondents supported terrorist attacks against soldiers and Jewish settlers, and 53 percent supported continued terrorist attacks against Israeli civilians inside the Green Line. At the same time, 84 percent supported a mutual cessation of violence, and 70 percent expressed support for an agreement between the organizations for a suspension of terrorist attacks.

7   Regarding the strengthening of the younger generation's influence and the undermining of the older generation's leadership, see Khalil Shikaki, "Palestinian Public Opinion and the *al Aqsa* Intifada," *Strategic Assessment* 5, no. 1 (2002): 15–20.

8   According to a public opinion survey published by *PSR – Survey Research Unit* in July 2001, support for Arafat plummeted from 46 percent in a survey published in July 2000, to 33 percent. Eighty-three percent of the survey respondents believed that the PA institutions were tainted with corruption, and 57 percent believed that corruption would increase in the future, or remain at the same level. The criticism of the PA is at least part of the reason for increasing support for the Islamic trend, from 17 percent in July 2000 to 27 percent in July 2001, paralleled by a drop in support for Fatah, from 37 percent to 29 percent. The survey published in December 2001 reflected similar trends, with slight differences, while the survey published in May 2002 indicated the emergence of a change in the attitude toward the PA and the character of the struggle against Israel. Regarding the PA, 83 percent expressed the belief that its institutions were corrupt, 91 percent supported substantial changes in the regime, 83 percent supported the holding of elections, and 48 percent supported (and 44 percent opposed) the transfer of authority from Arafat to a prime minister. The 19 percent support for Tanzim commander Marwan Barghouti was evidence of the increasing influence of grass roots forces. Barghouti was second in popularity to Arafat (35 percent support) at that time, while Hamas leader Ahmed Yassin was third with 13 percent support. The critical trend regarding the PA continued, as indicated by the

results of a survey published in August 2002, in which 84 percent of the respondents preferred substantial reform in the PA and 69 percent expressed support for the appointment of a prime minister, but only 25 percent expressed confidence in the ability of a new government to carry out administrative reform and deal with Israel. Thirty-four percent believed that the new cabinet would implement the planned reform. The institute that conducted the survey attributed the declining support for reform in the PA to opposition to pressure from the United States, which was interpreted more as concern for the security of Israel than as interest in democracy and the introduction of proper administration in the territories.

9   In August 2000, Fayyad founded the Palestine Investments Fund, designed to amalgamate all the PA's investments and transactions around the world. The PA's investments were estimated at $600 million. Fuel, cement, and other monopolies remained outside the purview of Fayyad's ministry, and were a key element in the accusations of corruption leveled by the opposition against senior PA officials (*Globes*, March 3, 2003).

10 The annual budget of Hamas is estimated at $10–14 million. Its main source of financing is the Saudi Arabian government and Israeli Arab citizens (*Ha'aretz*, September 18, 2003; *Yediot Ahronot*, August 1, 2003).

11 Fifty-six percent of those questioned in a survey published in 2002 by the *PSR – Survey Research Unit* supported PA measures against attacks in Israel as part of a mutual ceasefire, and 73 percent said they believed that the peace process would be harmed unless the PA took action against the perpetrators of terrorist attacks against Israeli citizens within the Green Line after an agreement on a mutual ceasefire. Eighty-two percent of the respondents, however, expressed concern that such action would lead to internal Palestinian clashes.

12 A survey published in June 2003 by *PSR – Survey Research Unit* showed diminished expectations of the Abu Mazen government, compared with the previous survey, published in April 2003. Support fell from 61 percent to 52 percent, confidence in his ability to carry out reform in the regime fell from 43 percent to 38 percent, and confidence in his ability to control the security situation fell from 39 percent to 35 percent. Thirty-six percent of the respondents expressed confidence that Arafat was more capable than Abu Mazen of reaching a political settlement with Israel, while 21 percent believed that Abu Mazen's chances were better than those of Arafat.

13 According to a survey of Palestinian public opinion conducted by the Truman Institute and *PSR*, 80 percent of the respondents supported an unlimited halt in violence, and 73 percent supported a one year ceasefire between the Palestinian organizations, during which neither side would use weapons (reported in *Yediot Ahronot*, July 2, 2003).

14 According to a *PSR* survey in October 2003, support for Arafat rose at that time to the highest level in five years. This can be interpreted as a response to the decision in principle by the Israeli cabinet to expel him. In addition, 85 percent of the respondents expressed support for the *hudna*, and 60 percent preferred a union of the various security forces under Arafat's command.

15 JCSS Annual Survey on Public Opinion and National Security, 2004 (press release, March 14, 2004): Fifty-six percent of the respondents supported disengagement from the Palestinians, but support fell to 50 percent if dismantling of settlements was involved. Sixty-eight percent believed that disengagement would reduce the risk of suicide terrorist attacks, and 62 percent believed that disengagement would improve security. Sixty-one percent believed that disengagement would improve Israel's economic situation, and 60 percent regarded it as a means of ensuring a Jewish majority in Israel. Sixty-two percent nevertheless believed that disengagement would constitute withdrawal under fire, demonstrating weakness, and 55 percent that it would amount to giving a prize for terrorism. Eighty percent of respondents supported the security fence, 60 percent supported including

extensive areas in the West Bank on the Israeli side of the security fence, although not isolated settlements. Seventy-one percent expressed opposition to a fence along the Green Line that would leave the settlement blocs outside. According to a survey conducted by *PSR* (*Jerusalem Post*, March 28, 2004), 58 percent of Palestinian respondents supported a coordinated Israeli withdrawal, while 38 percent preferred a unilateral Israeli withdrawal. Thirty-two percent of respondents expressed concern that Israeli disengagement from the Gaza Strip would strengthen Israel's hold in the West Bank, and foster an internal Palestinian confrontation. Forty-nine percent believed that withdrawal would lead to a reduction in terrorist attacks from Gaza. Fifty-eight percent of respondents believed that the separation fence would lead to an increase in terrorist attacks. According to a survey conducted by the Truman Institute and *PSR* (*Yediot Ahronot*, April 5, 2004), 64 percent of Israeli respondents supported the disengagement plan, although 60 percent preferred a negotiated agreement to unilateral disengagement.

CHAPTER ➤➤➤

9

The 2004 Strategic Balance:
A Net Assessment

MEIR ELRAN

The history of the Middle East has been long been punctuated by dramatic changes and unexpected crises. As such, the region presents a difficult challenge to those hoping to mold its future, be they local forces or external elements. Researchers who attempt to unveil the underlying currents of the Middle East and those who would extrapolate from the past and the present to suggest trends that may subsequently unfold encounter a similar difficulty in their efforts to tackle a region full of uncertainty with its own rapidly changing dynamics.

The year 2003 and the first half of 2004 were marked by major upheavals, significant even for a region that is no stranger to change and turmoil, and an already intricate regional picture has become even more complex. To the sensitive interstate texture has been added the occupying military presence of a superpower, along with a continued and even intensified process of proliferation of sub-national forces, which undermine the status quo of established governmental regimes. Hence the uphill challenge of constructing an orderly regional picture and forming a net assessment intended to provide a qualitative comparative evaluation that encompasses numerous elements, some of which are difficult to measure.

This chapter adopts the definition of "net assessment" used by the American defense establishment, namely, "a comparative analysis of technological, political, economic, and other factors governing the relative . . . capabilities of nations. Its purpose is to identify problems and opportunities that deserve the attention of decision makers and the interested public." Thus, the analysis that follows is comparative, focusing on some of the primary influences acting in the region. The picture that emerges intends to be integrative, presenting the major factors and evaluating their significance. In the assessment that follows, Israel constitutes the principal "interested public," and as such the picture depicts the latent dangers and opportunities largely from Israel's viewpoint.

The outstanding trademark of the Middle East of 2003–mid-2004 continued to be, as it has been for generations, a grave, deep-rooted, militant confrontation. The confrontation once again pitted the Arab countries against Israel, and a comprehensive peaceful solution to this conflict was still far off. Nonetheless, this traditional issue received less attention, as emphasis has shifted to the intensifying new political confrontation between opposing societies and cultures. On both the political and the ideological-cultural levels, the confrontation took on a violent form and was manifested through a wide variety of forces and alternative means. Concomitantly, the security element in all its dimensions (conventional, non-conventional, and sub-conventional) continued to represent a major focus of the region's players. Therefore, this issue necessarily assumes a key role in assessing the strategic balance in the Middle East.

With the evolution of the confrontation from the purely political interstate to the ideological arena, however, it is clear that despite the overriding importance of the security factor, the significance of other elements – though more amorphous in their composition and harder to evaluate – increased. The elements in question were political, social, economic, and related to image and self-awareness, and together and independently created a far more obscure picture than the conventional military one. In this complex situation, that which was manifested as strength frequently contained basic weaknesses, and vice versa. Apparently strong players found themselves restricted in the use of their power and their ability to impose their will on their opponents. Conversely, seemingly weak players found themselves capable of facing up to their opponents relatively successfully by making informed use of particular tools, even if such tools were unacceptable to the stronger rivals. In the Middle East, therefore, there were neither absolutely strong nor absolutely weak players. Evaluation of their relative strengths and weakness forms the central issue in this chapter.

The following net assessment is drawn from the material of the previous chapters and the quantitative data on the region's military forces. It focuses on four major sets of participants: the Arab countries and Iran; the sub-national Islamic and Arab actors, especially those active in the Palestinian theater; the international community, particularly the United States; and the State of Israel. Based on a multilateral evaluation of their weaknesses and strengths, the chapter assesses these groups and their influence on the Middle East balance and Israel's strategic environment.

➤ THE GENERAL STRATEGIC BALANCE

An aerial photograph of the Middle East of 2003–2004 might show an arena dominated by two major groups struggling intensively with one another, each woven from its own ideological, cultural, and political cloths. This arena was surrounded by many bystanders who supported one side or the other but did not actively participate in the confrontation taking place before them.

On one side of the arena was a group comprising forces propelled by radical Islamic ideology to battle the political and cultural status quo in the Middle East, including mainstream Arab regimes. Their ethos unequivocally rejected Western hegemony of any kind and behavior spawned by the stigmas of cultural, social, and economic inferiority to the prosperous, coercive Western world. The members of this group used violence to express their strategic concept, which stressed two principal, intertwined aims: opposition to external coercion; and the construction of a strong, viable alternative based on local culture and traditions and capable of defying the Western option, embodied particularly by the US. This non-homogeneous group included the international Islamic terrorist organizations associated with al-Qaeda; violent Palestinian groups, first and foremost Hamas; Hizbollah in Lebanon; domestic Iraqi forces resisting the American occupation; and Iran, the sole regional state striving to export the Islamic revolution.

This group was not a coordinated coalition. However, the common feature of its participants was clear: use of all means, including terrorism against civilians, to destroy the foundations of the enemy camp – the US and Israel. In 2003–2004 this group gained strength and acquired increasing support in the greater Muslim world and the Middle East in particular. Its rapid successful presentation of a major challenge to the US and the West was due to several important factors: the clarity and simplicity of its message; its ostensible authenticity in the growing Muslim region; a clear definition of the enemy using extreme, stereotyped images that inculcated hatred; the cultural-political-military proximity between the US and the already-abhorred Israel; determination to use uncompromising violence and unrestricted terrorism; emphasis of the value of sacrifice; and in addition to all these, the immense vulnerability of the Western world.

In this context Iran stood out as an important state in the region that associated itself ideologically with the anti-status quo group, actively supported various bodies among them, and affirmed its determined opposition to the US and Israel. However, its importance was enhanced to a great extent by its being the only country in the region, apart from Israel, developing a military nuclear option. Obviously the acquisition of nuclear weapons by Iran would significantly strengthen the radical camp. European diplomatic efforts in 2003–2004 to prevent Iran from attaining a nuclear option were somewhat successful and reduced the acute nature of the threat. Nevertheless, it seems that the Iranians have retained the potential to acquire a nuclear weapons capability and were guarding this potential tenaciously.

At the same time, the advantages enjoyed by the radical camp were severely constrained by several factors, chief among them the unmatched military resources and political power of the adversary. The American resolve to fight Islamic extremism using massive military strength and diplomatic influence was resonant of the determination displayed by Israel in its war against Palestinian terrorism. However, in mid-2004 backlash remained an unknown commodity, and it was not clear what kind of extremist sentiments and capabilities might be unleashed that were as yet dormant. Neither was it clear whether unbridled use of showcase terrorist attacks as the weapon of the weak against the strong would generate an aggressive response and puncture the civilian infrastructure of the radical camp. While the military response to terrorist provocation has been vehement, its effect in Iraq and among the Palestinians has so far been more to arouse public support for the radical elements than induce their rejection.

Dominating the other side of the arena were the US and Israel, which acted both separately and in tandem. The significant longstanding connection between the two countries, built on compatible if not shared political and cultural traditions, was enhanced by the US proactive response to the events of September 11 under the neoconservative auspices of President Bush's administration. This response reflected a transition from the traditional pattern of political involvement in order to influence the Middle East, to a combined use of political and military means in order to recreate the regional reality in accordance with American ideals and interests. The Western determination against radicalism was a principal ingredient in Middle East dynamics of 2003–2004 and overshadowed most of the region's local and interstate issues.

However, the very existence of the Israeli–American connection also hindered the US capacity to act in the region and recreate the political systems as per its own ideology. In the eyes of the Arab leaders – as well as the Muslim and Arab publics – this alliance was one of the principal obstacles preventing them from accepting or supporting the American vision. This ideological impediment joined another and no less important one, namely, the restrictions on the use of force by both the US and Israel. Although the understanding between the allies worked to Israel's advantage and enabled it to expand significantly its freedom of military action against the Palestinians, both powers were still prevented – mainly because of their political culture and democratic nature – from exploiting their full strength and military advantages in the war against civilian sources of resistance.

The confrontation between the two camps was intense and likely to last for some time, accompanied by respective achievements and losses. The complex fabric of weaknesses and strengths made it improbable that the struggle would be clearly decided on behalf of one side or another. For precisely this reason popular support was a critical ingredient for long-term endurance on both sides, notwithstanding the diametrically opposed nature of their host societies: thriving Western democracies that formed stable, well-ordered countries versus non-democratic societies guided mainly by religious precepts and whose socio-economic situation lay at the base of the global pyramid. Engaging public support was radically different in both camps, as were the sets of values around which the public was supposed to rally.

In the battle for public support among the respective combatants the principal groups were dominated by particular ideologies and behaviors. The violent hatred of the US and Israel and that which they represent was the major fuel in the radical camp's struggle against them. It was discernible in all Arab countries and in Iran, but assumed a more public and extremist nature where no local government exerted a restraining effect. It seemed that the deep assimilation of radical Islamic values coupled with anti-Western, anti-American, and anti-Israeli political consciousness created a strong and dangerous foundation of determination to stand up and fight against the "Satanic enemies," even at a high and long-term cost. Hence also the volatility of occupied Iraq, which presented a potential for conflagration that was liable to cause chaos to no less of an extent than in the Palestinian territories.

In Israel, after nearly four years of violence, with ongoing acts of terrorism and a high number of casualties among civilians both relative to previous wars and to the size of the population, the Israeli public demonstrated a capacity to stand firm. The majority of the Jewish public in Israel supported the struggle against Palestinian violence, and steadfastness seemed likely to continue if the public remained convinced that this was primarily a defensive war against terrorism. It also appeared that as long as the Israeli government presented to the public at least the semblance of a diplomatic track, coupled with ongoing general American support, it would continue to retain public approval for the war against terrorism, the price notwithstanding. This clearly indicated a failure by the Palestinians to wield terrorism as an agent for public pressure on the Israeli government to change its policies.

In contrast to the success of Israel and the radical Islamic camps in recruiting their respective publics for a long and hard struggle, the American public was a more unknown factor. The US began to confront ebbing public determination to lend long-term support for the aggressive neoconservative ideology and its forceful imposition on the Middle East. The deteriorating security situation in Iraq and the few successes in the total war against international terrorism were liable to weaken the public's determination further. Although the elections in November 2004 seemed unlikely to be decided solely according to Middle East and international terrorism issues, it was possible that a new administration would encourage fresh thinking and perhaps changes in American policy. This potential was clear to international terrorist organizations, to centers of resistance in Iraq, and to other bodies in the anti-status quo camp, who sought an American version of the Spanish volte face regarding Iraqi involvement.

Herein lay the two opposing bastions of the Middle East theater. Significant supporting actors on the international and regional stages followed the active struggle between the two camps closely. Their outstanding feature, however, was their primary status as observers. Indeed the EU, the UN, and Arab countries generally attempted to abstain from any initiative or active participation in the conflict arena. European pressure on Iran and the attempted promotion of the roadmap were notable exceptions to this passivity. For the most part, however, the Europeans opted to be bystanders, largely because of the lessons learned from the trans-Atlantic crisis surrounding the war in Iraq. They sat firmly and consistently on the fence, straddling their cultural-economic association with the West and their tendency, for a variety of considerations, to avoid direct confrontations with the radical camp.

The Arab countries, and mainly those defined as linked to the US, faced a strong dilemma. On the one hand they regarded radical Islam and extremist terrorism as presenting clear threats to their own regimes, and they preferred to see these alternatives snuffed out. On the other hand, identification by the Arab regimes with the American vision was untenable, since it presented an antithesis of their oligarchic rule and thus an existential threat to their own dictatorships, which were rooted in the poverty and political powerlessness of the Arab public. Similar objections targeted the American operations in the Middle East and US political and military support for Israel in its confrontation with the Palestinians. All these factors made the Arab countries as well as the European community less relevant and left the primary arena to the parties actually fighting there.

➤ THE MILITARY BALANCE

➤ The Conventional Balance

The year 2003 marked a significant and apparently long-term change in the conventional balance of forces in the Middle East. The conquest of Iraq by the US and its allies removed Iraq from the group of countries that could potentially form an Arab military coalition fighting against Israel. To be sure, over more than two decades – perhaps since the period of its war with Iran, its defeat in the Gulf War, and the loss of its major source of arms, the former Soviet Union – Iraq did not present a significant threat to Israel. Still, this long and gradual process of declining military strength was not pronounced enough to eliminate Iraq completely from what has been variously referred to since 1948 (though in retrospect perhaps exaggeratedly) as the "eastern front." This threat was consistently one of the important foundations of the Israeli defense concept and carried much weight in the buildup of Israel's military forces. As long as Iraq remains under American control or influence in one form or another, or alternatively is occupied with serious internal struggles regarding the form and structure of its regime as well as its economic and social rehabilitation, it may be assumed that its military potential against Israel will be negligible. In the absence of Iraq, there is no eastern front against Israel. From the conventional military aspect this development has far-reaching strategic significance, similar to that of the peace agreement with Egypt.

Since Egypt and Jordan have signed peace agreements with Israel and Iraq has been demilitarized, Israel no longer faces, at least for the foreseeable future, the threat of an Arab coalition fighting against it. The sole Arab country potentially presenting a threat against Israel is Syria. However, if the past is any indication, Syria would not initiate overall acts of aggression against Israel by itself and in the absence of clear support by additional Arab countries. Thus, Israel is free, at least in the immediate future, of a direct military threat against it.

This assessment is strengthened by complementary observations on the region at large. Syria, for example, continued to stagnate from a military point of view because of its economic incapability to purchase advanced weapon systems and its lack of backing by a major power. Such conditions inevitably spell inferiority, and certainly relative to its major enemy, Israel, which continued to build up its military strength. The Syrian army was presumably aware of its limitations when facing Israel and was consequently trying to find ways, while maintaining the defensive posture of its limited army, to exploit its few advantages as much as possible despite the threat of a direct Israeli attack against Syrian territory should it be provoked.

As Shlomo Brom and Yiftah Shapir showed in a previous chapter, Egypt continued the gradual process of building up its military strength, mainly as a result of its defense relations with the US. Most significant was the major weapon systems it purchased. However, despite this important increase in strength, the Egyptian army still struggled to break free of the traditional military thinking that characterized its long period of association with the Soviet Union. The Egyptian army, at least to date, has not adopted an updated military doctrine based on Western principles of warfare, suitable for its new weapon systems. This basic situation reduced the army's capabilities of exploiting the many advantages of the American weapon systems it possessed, and relegated it to an inferior position relative to Israel.

In the remaining Arab countries that surround Israel or those of the outlying circles, no significant changes have taken place in the last two years in the buildup of their conventional military forces that may narrow the qualitative gap between them and the IDF. In most issues this gap has even widened, largely because the IDF was endowed with an up-to-date military capability that reflected the successful fundamental transition to implement the principles of the Revolution in Military Affairs (RMA). This revolution, partially completed in the IDF, reflects the optimal exploitation of combat principles and elements based on information superiority, command and control capability, effective allocation of targets to the fire-producing units, precision strike capability, and high immunity for the platforms launching the precision guided munitions. These capabilities are designed to permit great attrition of the enemy forces and the destruction of selected targets in a relatively short time, while exploiting the latent synergistic potential in the maximum implementation of "jointness" of combined forces.

In 2004, very few armed forces in the world possessed all or some of the RMA capabilities that were spectacularly demonstrated by the US army in the 2003 military campaign that caused the rapid collapse of the Iraqi army. It appears that Israel lies in second place in the process of transition as per this revolution, and considerably ahead of advanced Western countries such as Britain, Germany, and France. Third World countries, including the Arab countries, were experiencing difficulties in entering these fields, and there were signs of their own doubts if they could overtake Israel in this arena. Meantime, the gap between them and Israel in this vital field widened, with little sign of a reversal. The Arab states' armies were rooted in traditional and increasingly outmoded systems for several reasons. One factor was the high cost of transition from a classic army to one possessing RMA capabilities. It requires personnel trained in high level technology, currently lacking in Arab states. Furthermore, there was a difficulty in obtaining the required technologies. Although the US was ready to supply its allies – Egypt and the Gulf States, for example – with advanced weapons, it made sure to maintain a gap between its own capabilities and those of other countries. In the case of the Middle East, the US also maintained the gap between Israel's capabilities and those of other countries in the region. Certain states, mainly Syria and Iran, were cut off from any source of the most advanced weapon systems.

While Israel's neighbors lacked important technological skills, Israel had its own technological and industrial capabilities, which permitted it to develop and produce those items that it could not purchase from foreign sources, for example in the field of space. The need to maintain information superiority and to enhance command and control capability, particularly in the case of distant opponents, obligates use of the field of space. Israel has already demonstrated the capability of developing intelligence and telecommunications satellites and it also has the proven capability of launching satellites. No other country in the Middle East had such a capability yet, and it was not expected that these gaps would close in coming years.

Quantitative gaps still existed to Israel's disadvantage relative to Arab armies regarding traditional platforms, such as tanks, artillery, and fighter aircraft. However, if we take into account realistic assessments of coalitions that can theoretically confront Israel, especially in the light of the results of the war in Iraq, then the accumulated Arab quantitative advantage is no longer as great and possibly not even significant. In the past Israel has succeeded in overcoming wide gaps even when confronting its enemies in the same field because of its various qualitative advantages, mainly superior manpower and experienced and advanced commanders. If the new capabilities and their influence on the battlefield are added, it may be assumed that the confrontation using conventional weapons systems would occur against an enemy whose combat capability has been seriously damaged. It was estimated that the results of such a confrontation would reflect the marked comprehensive military advantage enjoyed by Israel.

Overall, then, in a confrontation between the large regular armies of the Middle East, Israel would enjoy a definite advantage, which is likely to increase in the coming years. This is of great fundamental deterrent value. There is increasing awareness among the Arab leadership that an overall military confrontation would not supply them with a suitable environment for advancing their strategic objectives against Israel. This has positive ramifications regarding the degree of interstate stability in the region.

Yet with all its importance, this same Israeli military advantage has limited significance in the current era. This was mainly because Israel, whose major strategic objective is defensive, has found itself in a situation in which its high military investments over many years have borne fruit during a period in which the capabilities – and perhaps also the motivation – of the Arab countries to attack it were very low. The widespread recognition among the Arab leaderships that Israel's military strength was a given fact caused those Arab countries that wished to change the status quo (perhaps Syria, for example) to attempt to locate other means of realizing their interests, and not by conventional military confrontation between countries. On the other hand, should Israel wish to exploit its military superiority by attacking Arab countries, it would find itself restricted from acting on a broad front because of grave international constraints.

The Non-Conventional Balance

Israel's conventional military advantage is matched by a strong quantitative and qualitative advantage in the non-conventional arena. Libya and Iraq no longer figured on the list of countries threatening Israel with non-conventional weapons. Although apparently Iraqi non-conventional capability was doubtful even earlier, the uncertainty played to Iraq's advantage, as evidenced by the preparations in Israel against non-conventional weapons attacks on population centers. Syria still presented a certain residual threat, and Iran's nuclear capability, while not definitively known but clearly embodying serious potential, loomed as a serious threat.

It has been widely presumed that Israel possesses a highly developed nuclear arsenal and not insignificant capabilities in other fields of non-conventional weapons. Israel also has a very good capability of delivering its strategic armaments to distant targets by means of its air force, which enjoys decisive air superiority, and a developed array of missiles that could reach potential targets in neighboring states. In contrast, at this stage Israel's opponents have only chemical weapons and limited quantities of biological weapons. The air forces of the neighboring states have a circumscribed if real capability of delivering the weapons to targets deep inside Israeli territory, relying mainly on ground-to-ground missiles inferior to Israel's missiles but which could still inflict damage on the Israeli home front.

Israel was also the only Middle East country possessing a developed defense system against some of these threats. The Israeli defensive capability is multi-tiered and consists of passive defense measures, such as an early warning system, gas masks, sealed rooms, and medical facilities for the entire population; operational defense systems including the Wall and Patriot missile defense systems; and a capability under development to attack the weapons of mass destruction and their launching devices. Together the offensive and defensive superiority have lent Israel a not insignificant deterrent capability.

This advantage, however, by no means obliterates the threats directed to Israel and certain inherent weaknesses. Potentially hostile neighboring states possess a large number of missiles, albeit of inferior quality. This affords them a certain advantage, particularly in an engagement between missiles with conventional warheads. As a geographically small country in which most of the population is concentrated in an even smaller area, Israel's vulnerability is considerably increased – although this weakness is somewhat balanced by the fact that in several neighboring states, a high percentage of the population and the infrastructures are also concentrated in a small number of large cities. Furthermore, a conflict that targets civilians specifically poses a particular challenge to social resilience. The experience of the past, both during the 1991 Gulf War and during the preparations on the Israeli home front for the 2003 Iraq War, questions the degree of confidence that Israel's government attributes to its defense capabilities and the resistance of its population to a missile attack.

➤ THE SUB-CONVENTIONAL SECURITY BALANCE

It was precisely because Israel enjoyed such a decided advantage in military and non-conventional capability that the conflict assumed important new dimensions and evolved largely into a theater of sub-state low intensity conflict. Such confrontations generally signal the limitations of a global power, such as the US, or a local power, such as Israel, to wield its military strength and impose its strategic vision on forces actively challenging its superiority. In this context there was a growing resemblance between Israel's ongoing confrontation with Palestinian forces and that of the US struggle with Iraqi forces. Indeed, nearly four years into the current armed conflict with the Palestinians and one year after the fall of Baghdad, one conflict resonated of another, at least at the level of awareness and encouragement, if not yet at the level of active cooperation. Moreover, many in the Arab world regarded the war in both cases as being waged against the same American-Israeli coalition.

The Palestinian–Israeli confrontation was the older of the two Middle East conflicts currently underway by sub-state groups against a state power. In marked contrast to the conditions underlying the conventional balance, the highly charged political, cultural, and religious environment generated important momentum for the Palestinians and endowed them with capabilities disproportionate to their military assets. Perhaps chief among these was the very nature of the confrontation as a popular struggle. As such, critical to the ongoing violence was support of the Palestinian population at large and its capacity for continued resistance. The numerous varied organizations and groups actively engaging in terrorist attacks received broad support from the public, which regarded them as the spearhead of the war against the Israeli occupation. For its part, Israel has failed to drive a wedge between the terrorist activists and the Palestinian public. On the contrary, a significant number of Israel's military and diplomatic moves have strengthened the Palestinian public's support of its fighters. According to polls conducted by Palestinian research institutes, the Palestinian public in general has demonstrated consistent endurance *vis-à-vis* the confrontation with Israel and a steadfast belief in the justice of its continued fight. The feeling in the Palestinian street was that they have nothing to lose.

The Palestinian elements engaged in terrorism and guerilla activities could successfully assume a low profile because of their assimilation in the supportive civilian population. They formed an elusive target, making it difficult for Israel to locate them and thwart their operations. Furthermore, the Palestinian front was characterized by heavy activity, at least in its attempts to perpetrate significant attacks, limited only by

153

the Israeli efforts to obstruct these attempts. The fact that Palestinian terrorism was regarded as in perpetual strike mode cast the Israelis as constantly on the defensive.

In contrast to the IDF, which combated terrorism with well-trained enlisted troops equipped with advanced weapons, the Palestinian terrorists have scored significant achievements with steadily improving inferior weapons. The high availability of "platforms" for launching explosives, in the shape of suicide bombers of both sexes and fairly primitive weapons, created a significant reserve, permitting long-term planning of activities. Similarly, there was decided asymmetry between military and intelligence capabilities. Along with impressive Israeli capabilities, the Palestinians have also accrued significant intelligence capabilities for the location, identification, and assessment of targets. Moreover, the Palestinian capabilities have improved in this area, thus permitting them to exploit more precisely the means at their disposal. "Suitable" Israeli targets presented themselves to terrorists with relative ease, particularly in populated towns and cities. Thus, even when suicide bombings were executed with relatively low frequency, their impact was significant. Although particularly from the middle of 2003 and onward there has been a noticeable drop in the number of "successful" showcase attacks, infrequent showcase attacks still played a significant role in the Israeli–Palestinian confrontation.

Despite Israel's considerable and largely successful efforts to seal the Palestinian areas, there was steady input from foreign sources in the form of weapons as well as money for financing the terrorists and strengthening their civilian infrastructure. Hence the cooperation, albeit limited because of Israel's preventative measures, with foreign organizations and particularly with Hizbollah, supported and encouraged by Iran. On a more local level there was a discernible trend by terrorist organizations to include Israeli Arabs in the war effort against Israel, both for propaganda purposes and for the advantage they afforded through their freedom of movement in Israel. The efforts so far to enlist the Israeli Arab population met with limited success, but from the Palestinian standpoint, the Israeli Arab sector forms an important reservoir for recruits to the armed struggle.

Notwithstanding the difficulty in utilizing conventional resources to achieve a victory on this battlefield, the Israelis have benefited from their military assets. These include high quality intelligence in the form of human intelligence as well as innovative technological capabilities using Sigint and visual means. These highly developed capabilities formed an important tool both for thwarting terrorist attacks en route and for hitting point targets. Also critical was the presence of numerous ground forces, which included the optimal use of special forces, as well as an air and, in the case of the Gaza Strip, sea presence. The IDF had limited forces to deploy for protracted periods in the occupied territories for defensive assignments or for thwarting terrorist attacks, but relative to the Palestinian forces and based on precise intelligence, the very broad deployment of the forces was of great operational importance.

More than in previous conflicts, the fight against Palestinian terrorism in both the defensive and offensive modes has created a high degree of operational and tactical cooperation between the various security forces, especially between the IDF and the General Security Services (GSS), and within the various branches of the IDF. The operational cooperation between intelligence and the air force was outstanding, as was demonstrated particularly in targeted assassinations. In this context, the IDF's advanced technology, most notably its use of precision guided munitions, afforded a critical asset. The technological capability of focusing repeated attacks on specific objects and thus, generally successfully, avoiding extensive injury to innocent civilians, was of great importance in the struggle against the Palestinians. This included attacks on the Palestinian leadership, which has become a prominent part of the campaign. The express aim was to hit specific targets in the terrorist leadership, but clearly it had broader objectives as well, such as attacking Palestinian morale. Complementary diplomatic efforts against Palestinian leaders weakened the Palestinian political organization and limited leaders' freedom of international political action. Moreover, even if Israel considered its military activity in the territories as defensive, it was understood as punitive tactics, designed to deter the population from offering support and refuge to terrorists.

To these military edges should be added the country's passive defense system. The security fence, scheduled for completion by the end of 2005, is expected to play a central role in the defense against Palestinian terrorism. The fence is matched by an extensive military, police, and civilian security organization, deployed throughout the region, from the occupied territories to coffee houses in the heart of Israel's cities. In addition, certainly relative to the Palestinians but even in absolute terms, Israel continued to invest major sums of money in the war against terrorism.

Thus, nearly four years of the current armed Palestinian struggle have not resulted in an outright Israeli victory. Although the last year has seen a decline in the scale and frequency of showcase attacks, there is no evidence of a reduction in motivation and efforts by Palestinian terrorist organizations to execute such attacks. Rather than deterrence, Israel's success in thwarting terrorist attacks has sown among the Palestinians the seeds of continued hatred and a drive to intensify the long-term armed struggle against Israel. Certainly Israel has scored considerable successes in obstructing Palestinian terrorism, largely due to its military advantage. It is also possible that the latent potential of the new military capabilities and technologies to fight this low intensity confrontation more effectively have not yet been fully realized.

However, even should new military options be devised and adapted to this conflict, it is doubtful that this would significantly alter the ultimate picture. Since the criteria for victory are not necessarily military and the final outcome will not be determined on the battlefield, in many cases important achievements in destroying terrorist targets do not bring the desired results. Rather, the struggle will be decided, if at all, in other arenas: the creation and influence of political processes, the struggle for public opinion, the capacity to retain ideological motivation, and the preservation of popular strength and national resilience.

The US entanglement in Iraq, as different as it is from the Palestinian-Israeli arena, reinforces the conclusions as to the limitations of force employed by a superpower, however great its capabilities, in the face of forces driven by an ideology of cultural hatred and motivated to reject foreign coercion and foreign occupation.

➤ ISRAEL'S VANTAGE

Several major issues highlighted 2003–2004 as a period marking significant improvement to Israel's strategic situation. Chief among these was the extensive political and military support of the US administration for Israel, perhaps unmatched in the history of the excellent relations between the two countries. The ideological basis of this partnership awards it special strength, and the American support for the unilateral disengagement plan was a further important factor in this alliance. In addition, the dismantling of the Iraqi military and political organizations definitively canceled the threat of an eastern front as well as a viable Arab coalition. Israel's conventional and non-conventional military superiority in the region remained strong and reduced the threat of an offensive initiated by Syria or any neighboring state, unless there was a radical change of regime and attitude.

The war against Palestinian terrorism appeared to have no imminent end. Since the peak period in 2002, terrorism has decreased in frequency and magnitude, particularly from the second half of 2003. The construction of the security fence, despite the political, humanitarian, and diplomatic complications associated with it, continued and contributed significantly to the defense capabilities. Israel has also achieved political successes in the war against the Palestinians, and for the first time in a long period, has taken the diplomatic initiative with its unilateral disengagement plan. Israel became the party dictating the agenda, which has earned it much domestic and international approval. In contrast, the Palestinian Authority was steadily declining, both internally and in the international venue. The Palestinian political structure has been marked by impotence, disintegration, and retrenchment, which barred any opportunity for significant constructive action.

Furthermore, Israel's domestic situation has improved primarily with regard to the economy, which showed important signs of recovery after years of a serious recession. The economic and social gap in the Israel–Arab balance was still immense, which offered Israel considerable additional strategic advantages for the long term.

At the same time, the Israeli landscape was not free of difficulties and long-range problems. The intensifying Israeli–Palestinian conflict has assumed aggravated dimensions of deep mutual hatred dictated by the national narratives of the two sides. The concept of disengagement expressed in the Israeli discourse was apparently designed to separate the warring parties and place a partition between them. However, this may well be an illusion, if not a deliberate denial of the cruel, long-term reality. It was clear that the disengagement plan, at least in its present form, was not a substitute for resolving the conflict through a negotiated agreement. The basic conditions of the conflict exist and are getting worse. The social-demographic clock is ticking faster and, in the absence of a complete, agreed solution, which currently is not in sight, it will become a time bomb.

The most outstanding issue on Israel's domestic scene was the relationship between the Israeli Arab minority and the Jewish majority. Israel has enjoyed relative tranquility and stability notwithstanding the sensitive, complex nature of the relationship. However, in recent years exacerbated by the violent Israeli–Palestinian confrontation, new cracks have appeared that are liable to threaten the delicate fabric of the relations between these populations. The tension between the sectors compounds the demographic anxiety, which here too is a source of concern if a Jewish majority is an essential condition for the preservation of the Jewish character of the State of Israel. Despite the growing tension and the riots of October 2000 that left thirteen Israeli Arabs dead, Israel has taking insufficient steps to ease this charged problem. If neglected further, the situation may take a violent form and encourage the recruitment of Israeli Arabs to the Palestinian terrorist cause.

On the international scene Israel is supported solely by the US. The international community has condemned Israel's policy toward the Palestinians in a manner arousing strong feelings of hostility that feeds, and is perhaps itself fed by, increasing anti-Semitism. So too in the regional context, where even the relations with Arab countries that have recognized Israel and signed peace agreements with it are at best formal and in fact are often driven by both official and public hostility. American support notwithstanding, Israel would benefit from greater international legitimacy and a more balanced international environment.

Thus, there is apparently a strategic balance between Israel and its neighbors, so that neither side is capable of overcoming the other. At the same time, a window of opportunity for Israel looms on the horizon. Its overall strength relative to its opponents allows it to overcome the difficulties it is currently facing, and its overall advantage in the strategic balance, even if not without important limitations, enables it to take considerable risks in order to create a better basis for the chance of diplomatic progress with the Palestinians. Ignoring this chance is liable to lead to great setbacks in the future, greater than those experienced by Israeli society in recent years.

Documents on Israel's Disengagement Plan

Ariel Sharon's Four-Stage Disengagement Plan,
May 28, 2004

I   BACKGROUND – DIPLOMATIC AND SECURITY SIGNIFICANCE

The State of Israel is committed to the peace process and endeavors to reach an agreed arrangement based on the vision presented by U.S. President George W. Bush.

The State of Israel believes it must take action to improve the current situation. The State of Israel has reached the conclusion that there is currently no partner on the Palestinian side with whom progress can be made on a bilateral process. Given this, a four-stage disengagement plan has been drawn up, based on the following considerations:

A. The stalemate embodied in the current situation is damaging; in order to break the stalemate, the State of Israel must initiate a process that is not dependent on cooperation with the Palestinians.

B. The aim of the plan is to bring about a better security, diplomatic, economic and demographic reality.

C. In any future permanent arrangement, there will be no Israeli presence in the Gaza Strip. On the other hand, it is clear that some parts of Judea and Samaria (including key concentrations of Jewish settlements, civilian communities, security zones and areas in which Israel has a vested interest) will remain part of the State of Israel.

D. The State of Israel supports the efforts of the United States, which is working along with the international community, to promote the process of reform, the establishment of institutions and improving the economic and welfare conditions of the Palestinian people, so that a new Palestinian leadership can arise, capable of proving it can fulfill its obligations under the road map.

E. The withdrawal from the Gaza Strip and from the northern part of Samaria will reduce interaction with the Palestinian population.

F. Completion of the four-stage disengagement plan will negate any claims on Israel regarding its responsibility for the Palestinian population of the Gaza Strip.

G. The process of graduated disengagement does not detract from existing agreements between Israel and the Palestinians. The relevant security arrangements will remain in force.

H. International support for the four-stage disengagement plan is widespread and important. This support is vital in ensuring that the Palestinians fulfill their obligations in terms of fighting terror and implementing reforms, in accordance with the road map. Only then will the sides be able to resume negotiations.

II   KEY POINTS OF THE PLAN

A   The Gaza Strip

1. The State of Israel will withdraw from the Gaza Strip, including all Israeli settlements, and will redeploy outside the area of the Strip. The method of the withdrawal, with the exception of a military presence in the area adjacent to the border between Gaza and Egypt (the Philadelphi route), will be detailed below.

2. Once the move has been completed, there will be no permanent Israeli military presence in the evacuated territorial area of the Gaza Strip.

3. As a result of this, there will be no basis to the claim that the Strip is occupied land.

B   Judea and Samaria

1. The State of Israel will withdraw from northern Samaria (four settlements: Ganim, Kadim, Sa-Nur and Homesh) as well as all permanent military installations in the area, and will redeploy outside the evacuated area.

2. Once the move has been completed, there will be no permanent Israeli military presence in the area.

3. The move will provide Palestinian territorial contiguity in the northern parts of Samaria.

4. The State of Israel, along with the international community, will help improve the transportation infrastructure in Judea and Samaria, with the goal of providing continuous transport for Palestinians in Judea and Samaria.

5. The move will make it easier for Palestinians to live a normal life in Judea and Samaria, and will facilitate economic and commercial activity.

C   The process

The withdrawal process is slated to end by the end of 2005.

The settlements will be split into the following four groups:

1. Group A – Morag, Netzarim, Kfar Darom
2. Group B – The four settlements in northern Samaria (Ganim, Kadim, Sa-Nur and Homesh).
3. Group C – The Gush Katif bloc of settlements.
4. Group D – The settlements in the northern Gaza Strip (Alei Sinai, Dugit and Nissanit)

The necessary preparations will be undertaken in order to implement the four-stage disengagement plan (including administrative work to set relevant criteria, definitions and preparation of the necessary legislation.)

The government will discuss and decide separately on the evacuation of each of the above-mentioned groups.

D   The security fence

The State of Israel will continue to construct the security fence, in accordance with the relevant cabinet decisions. In deciding on the route of the fence, humanitarian considerations will be taken into account.

III   THE SECURITY REALITY AFTER THE EVACUATION

A   The Gaza Strip

1. The State of Israel will monitor and supervise the outer envelope on land, will have exclusive control of the Gaza airspace, and will continue its military activity along the Gaza Strip's coastline.

2. The Gaza Strip will be completely demilitarized of arms banned by current agreements between the sides.

3. The State of Israel reserves the basic right to self defense, which includes taking preventive measures as well as the use of force against threats originating in the Gaza Strip.

B   The West Bank

1. After the evacuation of the northern Samaria settlements, there will be no permanent military presence in that area.

2. The State of Israel reserves the basic right to self defense, which includes taking preventive measures as well as the use of force against threats originating in the area.

3. Military activity will remain in its current framework in the rest of the West Bank. The State of Israel will, if circumstances allow, consider reducing its activity in Palestinian cities.

4. The State of Israel will work to reduce the number of checkpoints throughout the West Bank

IV   MILITARY INFRASTRUCTURE AND INSTALLATIONS IN THE GAZA STRIP AND THE NORTHERN SAMARIA REGION

All will be dismantled and evacuated, except for those that the State of Israel decides to transfer to an authorized body.

V   THE NATURE OF THE SECURITY ASSISTANCE TO THE PALESTINIANS

The State of Israel agrees that in coordination with it, consulting, assistance and training will be provided to Palestinian security forces for the purpose of fighting terror and maintaining the public order. The assistance will be provided by American, British, Egyptian, Jordanian or other experts, as will be agreed upon with Israel.

The State of Israel stresses that it will not agree to any foreign security presence in Gaza or the West Bank without its consent.

VI THE BORDER AREA BETWEEN THE STRIP AND EGYPT (THE PHILADELPHI ROUTE)

The State of Israel will continue to maintain military presence along the border between the Gaza Strip and Egypt (the Philadelphi route.) This presence is an essential security requirement. The physical widening of the route where the military activity will take place, may be necessary in certain areas.

The possibility of evacuating the area will be considered later on. This evacuation would be conditioned, among other factors, on the security reality and on the level of cooperation by Egypt in creating an alternative credible arrangement.

If and when the conditions are met enabling the evacuation of the area, the State of Israel will be willing to consider the possibility of setting up an airport and a seaport in the Gaza Strip, subject to arrangements agreed upon with the State of Israel.

VII REAL ESTATE

In general, houses belonging to the settlers, and other sensitive structures such as synagogues will not be left behind. The State of Israel will aspire to transfer other structures, such as industrial and agricultural facilities, to an international third party that will use them for the benefit of the Palestinian population.

The Erez industrial zone will be transferred to an agreed-upon Palestinian or international body.

The State of Israel along with Egypt will examine the possibility of setting up a joint industrial zone on the border between Israel, Egypt and the Gaza Strip.

VIII INFRASTRUCTURE AND CIVILIAN ARRANGEMENTS

The water, electricity, sewage and communications infrastructures will be left in place.

As a rule, Israel will enable the continued supply of electricity, water, gas and fuel to the Palestinians, under the existing arrangements and full compensation.

The existing arrangements, including the arrangements with regard to water and the electromagnetic area, will remain valid.

IX THE ACTIVITY OF THE INTERNATIONAL CIVILIAN ORGANIZATIONS

The State of Israel views very favorably continued activity of the international humanitarian organizations and those that deal will civil development, which aid the Palestinian population.

The State of Israel will coordinate with the international organizations the arrangements that will make this activity easier.

The State of Israel suggests that an international mechanism (such as the AHLC) be set up, in coordination with Israel and international bodies, that will work to develop the Palestinian economy.

X   ECONOMIC ARRANGEMENTS

In general, the economic arrangements that are currently in effect between Israel and the Palestinians will remain valid. These arrangements include, among other things:

A. The movement of goods between the Gaza Strip, Judea and Samaria, Israel and foreign countries.

B. The monetary regime.

C. The taxation arrangements and the customs envelope.

D. Postal and communications arrangements.

E. The entry of workers into Israel in accordance with the existing criteria.

In the long run, and in accordance with the Israeli interest in encouraging Palestinian economic independence, The State of Israel aspires to reduce the number of Palestinian workers entering Israel, and eventually to completely stop their entrance. The State of Israel will support the development of employment sources in the Gaza Strip and in the Palestinian areas in the West Bank, by international bodies.

XI   THE INTERNATIONAL CROSSING POINTS

A   The international crossing point between the Gaza Strip and Egypt

1. The existing arrangements will remain in force.

2. Israel is interested in transferring the crossing point to the "border triangle," south of its current location. This will be done in coordination with the Egyptian government. This will allow the expansion of the hours of activity at the crossing point.

B   The international crossing points between Judea and Samaria, and Jordan

The existing arrangements will remain in force.

XII   THE EREZ CROSSING POINT

The Erez crossing point will be moved into the territory of the State of Israel according to a timetable that will be determined separately.

XIII SUMMARY

The implementation of the four-stage disengagement plan will bring about an improvement in the situation and a break from the current stagnation. If and when the Palestinian side shows a willingness, an ability and an implementation of actions to fight terrorism, a full cessation of terror and violence and the carrying out of reforms according to the roadmap, it will be possible to return to the track of discussions and negotiations.

President Bush's Letter to Israeli Prime Minister Sharon, April 14, 2004

His Excellency Ariel Sharon, Prime Minister of Israel,

Dear Mr. Prime Minister:

Thank you for your letter setting out your disengagement plan.

The United States remains hopeful and determined to find a way forward toward a resolution of the Israeli–Palestinian dispute. I remain committed to my June 24, 2002 vision of two states living side by side in peace and security as the key to peace, and to the roadmap as the route to get there.

We welcome the disengagement plan you have prepared, under which Israel would withdraw certain military installations and all settlements from Gaza, and withdraw certain military installations and settlements in the West Bank. These steps described in the plan will mark real progress toward realizing my June 24, 2002 vision, and make a real contribution towards peace. We also understand that, in this context, Israel believes it is important to bring new opportunities to the Negev and the Galilee. We are hopeful that steps pursuant to this plan, consistent with my vision, will remind all states and parties of their own obligations under the roadmap.

The United States appreciates the risks such an undertaking represents. I therefore want to reassure you on several points.

First, the United States remains committed to my vision and to its implementation as described in the roadmap. The United States will do its utmost to prevent any attempt by anyone to impose any other plan. Under the roadmap, Palestinians must undertake an immediate cessation of armed activity and all acts of violence against Israelis anywhere, and all official Palestinian institutions must end incitement against Israel. The Palestinian leadership must act decisively against terror, including sustained, targeted, and effective operations to stop terrorism and dismantle terrorist capabilities and infrastructure. Palestinians must undertake a comprehensive and fundamental political reform that includes a strong parliamentary democracy and an empowered prime minister.

Second, there will be no security for Israelis or Palestinians until they and all states, in the region and beyond, join together to fight terrorism and dismantle terrorist organizations. The United States reiterates its steadfast commitment to Israel's security, including secure, defensible borders, and to preserve and strengthen Israel's capability to deter and defend itself, by itself, against any threat or possible combination of threats.

Third, Israel will retain its right to defend itself against terrorism, including to take actions against terrorist organizations. The United States will lead efforts, working together with Jordan, Egypt, and others in the international community, to build the capacity and will of Palestinian institutions to fight terrorism, dismantle terrorist organizations, and prevent the areas from which Israel has withdrawn from posing a threat that would have to be addressed by any other means. The United States understands that after Israel withdraws from Gaza and/or parts of the West Bank, and pending agreements on other arrangements, existing arrangements regarding control of airspace, territorial waters, and land passages of the West Bank and Gaza will continue. The United States is strongly committed to Israel's security and well-being as a Jewish state. It seems clear that an agreed, just, fair, and realistic framework for a solution to the Palestinian refugee issue as part of any final status agreement will need to be found through the establishment of a Palestinian state, and the settling of Palestinian refugees there, rather than in Israel.

As part of a final peace settlement, Israel must have secure and recognized borders, which should emerge from negotiations between the parties in accordance with UNSC Resolutions 242 and 338. In light of new realities on the ground, including already existing major Israeli populations centers, it is unrealistic to expect that the outcome of final status negotiations will be a full and complete return to the armistice lines of 1949, and all previous efforts to negotiate a two-state solution have reached the same conclusion. It is realistic to expect that any final status agreement will only be achieved on the basis of mutually agreed changes that reflect these realities.

I know that, as you state in your letter, you are aware that certain responsibilities face the State of Israel. Among these, your government has stated that the barrier being erected by Israel should be a security rather than political barrier, should be temporary rather than permanent, and therefore not prejudice any final status issues including final borders, and its route should take into account, consistent with security needs, its impact on Palestinians not engaged in terrorist activities.

As you know, the United States supports the establishment of a Palestinian state that is viable, contiguous, sovereign, and independent, so that the Palestinian people can build their own future in accordance with my vision set forth in June 2002 and with the path set forth in the roadmap. The United States will join with others in the international community to foster the development of democratic political institutions and new leadership committed to those institutions, the reconstruction of civic institutions, the growth of a free and prosperous economy, and the building of capable security institutions dedicated to maintaining law and order and dismantling terrorist organizations.

A peace settlement negotiated between Israelis and Palestinians would be a great boon not only to those peoples but to the peoples of the entire region. Accordingly, the United States believes that all states in the region have special responsibilities: to support the building of the institutions of a Palestinian state; to fight terrorism, and cut off all forms of assistance to individuals and groups engaged in terrorism; and to begin now to move toward more normal relations with the State of Israel. These actions would be true contributions to building peace in the region.

Mr. Prime Minister, you have described a bold and historic initiative that can make an important contribution to peace. I commend your efforts and your courageous decision which I support. As a close friend and ally, the United States intends to work closely with you to help make it a success.

Sincerely,

GEORGE W. BUSH

Letter from Dov Weisglass to Dr. Condoleeza Rice

The following is the full text of a letter sent by Prime Minister Ariel Sharon's bureau chief, Dov Weisglass, to the U.S. National Security Adviser Condoleeza Rice laying out the understandings reached between Sharon and U.S. President George W. Bush during their meeting at the White House on April 14, 2004.

Dr. Condoleezza Rice
National Security Adviser
The White House
Washington, D.C.

Dear Dr. Rice,

On behalf of the Prime Minister of the State of Israel, Mr. Ariel Sharon, I wish to reconfirm the following understanding, which had been reached between us:

1. Restrictions on settlement growth: within the agreed principles of settlement activities, an effort will be made in the next few days to have a better definition of the construction line of settlements in Judea and Samaria [the West Bank]. An Israeli team, in conjunction with Ambassador Kurtzer, will review aerial photos of settlements and will jointly define the construction line of each of the settlements.

2. Removal of unauthorized outposts: the Prime Minister and the Minister of defense, jointly, will prepare a list of unauthorized outposts with indicative dates of their removal; the Israeli Defense forces and/or the Israeli Police will take continuous action to remove those outposts in the targeted dates. The said list will be presented to Ambassador Kurtzer within 30 days.

3. Mobility restrictions in Judea & Samaria: the Minister of Defense will provide Ambassador Kurtzer with a map indicating roadblocks and other transportational barriers posed across Judea & Samaria. A list of barriers already removed and a timetable for further removals will be included in this list. Needless to say, the matter of the existence of transportational barriers fully depends on the current security situation and might be changed accordingly.

4. Legal attachments of Palestinian revenues: the matter is pending in various courts of law in Israel, awaiting judicial decisions. We will urge the State Attorney's office to take any possible legal measure to expedite the rendering of those decisions.

5. The Government of Israel extends to the Government of the United States the following assurances:

a. The Israeli government remains committed to the two-state solution – Israel and Palestine living side by side in peace and security – as the key to peace in the Middle East.

b. The Israeli government remains committed to the Roadmap as the only route to achieving the two-state solution.

c. The Israeli government believes that its disengagement plan and related steps on the West Bank concerning settlement growth, unauthorized outposts, and easing of restrictions on the movement of Palestinians not engaged in terror are consistent with the Roadmap and, in many cases, are steps actually called for in certain phases of the Roadmap.

d. The Israeli government believes that further steps by it, even if consistent with the Roadmap, cannot be taken absent the emergence of a Palestinian partner committed to peace, democratic reform, and the fight against terror.

e. Once such a Palestinian partner emerges, the Israeli government will perform its obligations, as called for in the Roadmap, as part of the performance-based plan set out in the Roadmap for reaching a negotiated final status agreement.

f. The Israeli government remains committed to the negotiation between the parties of a final status resolution of all outstanding issues.

g. The Government of Israel supports the United States' efforts to reform the Palestinian security services to meet their roadmap obligations to fight terror. Israel also supports the American efforts, working with the international community, to promote the reform process, build institutions, and improve the economy of the Palestinian Authority and to enhance the welfare of its people, in the hope that a new Palestinian leadership will prove able to fulfill its obligations under the Roadmap. The Israeli Government will take all reasonable actions requested by these parties to facilitate these efforts.

h. As the Government of Israel has stated, the barrier being erected by Israel should be a security rather than a political barrier, should be temporary rather than permanent, and therefore not prejudice any final status issues including final borders, and its route should take into account, consistent with security needs, its impact on Palestinians not engaged in terrorist activities.

Sincerely,

Dov Weisglass
Chief of the Prime Minister's Bureau

PART ▶▶▶

II

Military Forces

Introductory Note

For each of the countries reviewed below, only total numbers of main weapon systems are given, without breakdown into further detail. General data on each country is presented, along with data on weapons of mass destruction and on space assets. The data on arms procurement as well as arms sales and military industry appears in a concise format. The material presented below is updated to May 2004.

The table representing the order-of-battle of each country often gives two numbers for each weapon category. The first number refers to quantities in active service, whereas the second number (in parentheses) refers to the total inventory.

Charts representing distribution of weapon systems in three distinct regions of the Middle East follow the review of the individual countries. The regions are:

1   Eastern Mediterranean (includes Egypt, Israel, Jordan, Lebanon, Syria, and Turkey)
2   The Persian Gulf (includes Bahrain, Iran, Iraq, Kuwait, Oman, Qatar, Saudi Arabia, and UAE)
3   North Africa and other countries (includes Algeria, Egypt, Libya, Morocco, and Tunisia. To these is added Sudan.)

Detailed data on the region's military forces is available online at the Jaffe Center website: <http://www.tau.ac.il/jcss/balance/index.html>.

Economic Data

The tables on economic data include data on GDP (in current US dollars) and defense expenditure only. Sources for the economic data are EIU Country Profiles, EIU Quarterly Reports, IMF International Financial Statistical Yearbook, and SIPRI Yearbook.

Data on military/defense expenditures in the Middle East is notoriously elusive. Hence it should be regarded primarily as an indication of procurement trends.

Arms Trade and Foreign Military Cooperation

Data on military acquisitions and sales as well as on security assistance and foreign military cooperation is limited to information pertaining to the past five years. The year in parentheses indicates the most recent information about the data.

Note on Symbols

The following symbols are used to denote instances where accurate data is not available:

NA	Data not available. This symbol is used in the economic data tables only.
~	The tilda is used in front of a number to denote an inexact number.
+	The weapon system is known to be in use, but the quantity is not known.

YIFTAH S. SHAPIR
JUNE 2004

Review of Armed Forces

1 ALGERIA

Major Changes

- Ahmed Ouyahia is Algeria's new prime minister.
- Over the past three years Algeria improved its security cooperation with the US. The most important new system supplied to Algeria by the US is the Beech 1900 reconnaissance aircraft, six of which are now in service. Also, American special forces are present in Algeria, training Algerian forces in counterterrorism tactics.
- At the same time, Russia remains Algeria's major arms supplier. Over the past three years Russia rescheduled Algeria's debt of $4 billion. It upgraded Algeria's Su-24 combat aircraft. Recently Russia and Algeria signed a deal for Mi-171 helicopters, all of which have already been supplied. Russia also upgraded two of Algeria's frigates and one corvette.
- Other important arms deals included the recent acquisition of Russian-made MiG-29 combat aircraft from Belarus.
- The Algerian armed forces are now absorbing their new Ecureuil EC-550 light helicopters.

General Data

Official Name of the State: Democratic and Popular Republic of Algeria
Head of State: President of the High State Council Abd al-Aziz Buteflika
Prime Minister: Ahmed Ouyahia
Minister of Defense: Nureddin Zarhouni
Chief of General Staff: Major General Muhammad Lamari
Commander of the Ground Forces: Major General Salih Ahmad Jaid
Commander of the Air Force: Brigadier General Muhammad Ibn Suleiman
Commander of Air Defense Force: Brigadier General Achour Laoudi
Commander of the Navy: Admiral Brahim Dadci

Area: 2,460,500 sq. km.
Population: 32,800,000

Economic Data (in US $billion)

	1999	2000	2001	2002	2003
GDP (current prices)	48.3	54.2	54.7	53.9	62.1
Defense expenditure	1.83	1.88	NA	2.1	NA

Major Arms Suppliers

Major arms suppliers are Russia, which recently supplied combat aircraft, utility helicopters, and naval missiles. Russia also upgraded major weapon systems supplied in the past by the Soviet Union. The Czech Republic supplied tanks and training aircraft.

The US is becoming a major arms supplier to Algeria. It supplied Algeria with C³I systems, electronic reconnaissance aircraft, ground radars, and financial aid. Other suppliers are Belarus, which sold Soviet-made combat aircraft; Ukraine, which sold MBTs; and South Africa, which upgraded attack helicopters and sold UAV systems.

Foreign Military Cooperation

Type	Details
Forces deployed abroad	Congo (MONUC), Ethiopia, and Eritrea (UNMEE) (2001)
Foreign forces	US (2004)
Joint maneuvers	Italy (2003), US maritime SAR and ASW exercises (2002)
Security agreements	France (2000), Iran (2002), Libya (2001), Russia (2001), South Africa (2000), Turkey (2003)

Defense Production

Patrol boats, trucks, and small arms.

Strategic Assets

NBC Capabilities

Nuclear capability
One 15 Mw nuclear reactor, probably upgraded to 40 Mw (from PRC) allegedly serves a clandestine nuclear weapons program; one 1 Mw nuclear research reactor (from Argentina); basic R&D; signatory to the NPT. Safeguards agreement with the IAEA in force. Signed and ratified the African Nuclear Weapon-Free Zone Treaty (Pelindaba Treaty).

Chemical weapons and protective equipment
No data on CW activities available.

Signed and ratified the CWC.

Biological weapons
No data on BW activities available.

Not a party to the BWC.

Space Assets

Model	Type	Notes
Satellites		
ALSAT-1	Research satellite	Earth monitoring for natural disasters
Future launch		
ALSAT-2	Remote sensing	

Armed Forces

Order-of-Battle

Year	1999	2000	2001	2002	2003
General data					
Personnel (regular)	127,000	127,000	127,000	127,000	127,000
Ground forces					
Divisions	5	5	5	5	5
Total number of brigades	26	26	26	26	26
Tanks	860 (1,060)	900 (1,100)	900 (1,100)	900 (1,100)	900 (1,100)
APCs/AFVs	1,930	1,930	2,110*	2,110	1,980 (2,080)
Artillery	900 (985)	900 (985)	900 (985)	900 (985)	900 (985)
(including MRLs)					
Air force					
Combat aircraft	187 (205)	187 (205)	184 (214)	228 (258)	228 (258)
Transport aircraft	39 (45)	39 (45)	41 (46)	41 (46)	40 (45)
Helicopters	114	114	133* (142)	131 (140)	174 (183)
Air defense forces					
Heavy SAM batteries	11	11	11	11	11
Medium SAM batteries	18	18	18	18	18
Light SAM launchers	78	78	78	78	78
Navy					
Combat vessels	29	26	26	26	26
Patrol craft	21	21	16	16	16

* Due to change in estimate.

Personnel

	Regular	Reserves	Total
Ground forces	107,000	150,000	257,000
Air force	14,000		14,000
Navy	6,000		6,000
Total	**127,000**	**150,000**	**277,000**
Paramilitary			
National Security Force	16,000		16,000
Republican Guards Brigade	1,200		1,200
Gendarmerie	24,000		24,000

2   BAHRAIN

Major Changes

- In the second year since the coalition's occupation of Iraq, Bahrain remains a major US ally in the region and is officially a non-NATO ally of the US. The US Navy is upgrading its headquarters and logistics center in Bahrain.
- No major changes were recorded in the Bahraini armed forces.

General Data

Official Name of the State: State of Bahrain
Head of State: Amir Shaykh Hamad bin Isa al-Khalifa
Prime Minister: Khalifa ibn Salman al-Khalifa
Minister of Defense: Lieutenant General Khalifa ibn Ahmad al-Khalifa
Commander in Chief of the Armed Forces: Salman bin Hamad al-Khalifa
Chief of Staff of the Bahraini Defense Forces: Major General Rashid bin Abdallah al-Khalifa
Commander of the Air Force: Hamad ibn Abdallah al-Khalifa
Commander of the Navy: Lieutenant Commander Yusuf al-Maluallah

Area: 620 sq. km.
Population: 700,000

Economic Data (in US $billion)

	1998	1999	2000	2001	2002
GDP (current prices)	6.2	6.6	7.6	7.5	7.7
Defense expenditure	0.295	0.327	0.321	0.329	0.335

Major Arms Suppliers

Major arms suppliers are the US, which supplied combat aircraft, helicopters, and tactical ballistic missiles, and the UK, which supplied transport and training aircraft. Bahrain also received air defense systems from Sweden.

Foreign Military Cooperation

Type	Details
Foreign forces	Coalition forces: since the end of the war in Iraq numbers of American forces deployed in Bahrain has been reduced considerably.
Forces deployed abroad	Saudi Arabia (part of GCC "Desert Shield" Rapid Deployment Force)
Joint maneuvers	Egypt (2001), GCC countries (2001), Jordan (2001), US (2003)
Security agreements	US, Britain, GCC countries

Strategic Assets

NBC Capabilities

Nuclear capability
No known nuclear activity.

Signatory to the NPT.

Chemical weapons and protective equipment
No known CW activities.

Party to the CWC.

Biological weapons
No known BW activities.

Party to the BWC.

Future procurement
GID-3 CW detection system.

Ballistic Missiles

Model	Launchers	Missiles	Since	Notes
ATACMS		30	2002	

Armed Forces

Order-of-Battle

Year	1999	2000	2001	2002	2003
General data					
Personnel (regular)	7,400	7,400	7,400	8,200	8,200
SSM launchers		9	9	9	9
Ground forces					
Total number of brigades	3	3	3	3	3
Number of battalions	7	7	7	7	7
Tanks	180	180	180	180	180
APCs/AFVs	277 (297)	277 (297)	277 (297)	277 (297)	277 (297)
Artillery (including MRLs)	48 (50)	48 (50)	48 (50)	48 (50)	48 (50)
Air force					
Combat aircraft	24	34	34	34	34
Transport aircraft	2	2	2	2	3
Helicopters	41	39 (41)	39 (41)	40 (42)	40 (42)
Air defense forces					
Heavy SAM batteries	1	1	1	1	1
Medium SAM batteries	2	2	2	2	2
Light SAM launchers	40	40	40	40	40
Navy					
Combat vessels	11	11	11	11	11
Patrol craft	21	21	21	21	21

Personnel

	Regular	Reserves	Total
Ground forces	6,000		6,000
Air force	1,500		1,500
Navy	700		700
Total	**8,200**		**8,200**
Paramilitary			
Coast Guard and National Guard	2,000		2,000

3 EGYPT

Major Changes

- Egyptian industry received 125 more kits for the assembly of the M1A1 Abrams MBT. This will increase the total number of M1A1 MBTs in Egypt to 880 tanks.
- The Egyptian military industry will produce an unspecified quantity of the 155mm GH 52 Patria towed artillery (with auxiliary power unit) under technology transfer agreement with Finland.
- The army ordered 26 MLRS launchers with the new extended range (ER) rockets. The army will also receive 201 upgraded M109A3 SP howitzers from US Army drawdown.
- The air force is absorbing its 24 new F-16D it received last year in the framework of the Peace Vector VI deal.
- The air force is also engaged in a project to upgrade its five Hawkeye AEW aircraft and will get another aircraft from the US. The first of these upgraded aircraft entered service previously this year.
- The air defense force acquired 18 new early warning radars from the US, and it is now absorbing its refurbished and upgraded S-125 (SA-3) SAMs that were transformed to a self-propelled configuration by Ukraine.

General Data

Official Name of the State: The Arab Republic of Egypt
Head of State: President Muhammad Husni Mubarak
Prime Minister: Ahmed Nadhif
Minister of Defense and Military Production: Field Marshal Muhammad Hussayn Tantawi
Chief of General Staff: Lieutenant General Hamdi Wahaba
Commander of the Air Force: Maj. Gen. Magdi Galal Sha'rawi
Commander of the Navy: Vice Admiral Ahmad Saber Salim

Area: 1,000,258 sq. km. (dispute with Sudan over "Halaib triangle" area)
Population: 71,700,000

Economic Data (in US $billion)

	1999	2000	2001	2002	2003
GDP (current prices)	90.5	97.9	90.4	84.8	68.7
Defense expenditure*	2.31	2.46	2.61	2.75	NA

* Published defense expenditure data apparently does not include annual $1.3 bn foreign military assistance from the US.

Major Arms Suppliers

The major arms supplier is the US (MBTs, MLRS, combat aircraft, attack helicopters, radars, combat vessels, advanced air force and naval armament). Other suppliers include Germany (missile patrol boats), North Korea (ballistic missiles), PRC (training aircraft), Netherlands (AIFVs), Ukraine (upgrading SAMs and tanks), Belarus (upgrading SAMs), Russia (upgrading SAMs).

Major Arms Transfers

Bosnia (tanks, artillery)

Foreign Military Cooperation

Type	Details
Foreign forces	US forces as of September 2003 include some 400 soldiers; MFO's soldiers as follows: Australia (25), Canada (29), Colombia (358), Fiji (338), France (15), Hungary (41), Italy (75), New Zealand (26), Norway (4), Uruguay (60), US (865)
Forces deployed abroad	Peninsula shield force (2003), Georgia (UNOMIG), East Timor (UNMISET), Western Sahara (MINURSO), Sierra Leone (UNAMSIL), Democratic Republic of the Congo (MONUC)
Joint maneuvers	France, GCC countries (2001), Germany (2001), Greece (2001), Italy (2002), Jordan (2001), Netherlands, Saudi Arabia (2003), Spain (2001), UK (2001), US (2001)

Defense Production

Ballistic missiles, assembly of American MBTs, artillery pieces. Upgrading of AFVs. Assembly of basic training aircraft. Small patrol boats. Electronics and optronic equipment.

Strategic Assets

NBC Capabilities

Nuclear capability
22 Mw research reactor from Argentina, completed 1997; 2 Mw research reactor from the USSR, in operation since 1961.

Party to the NPT. Safeguards agreement with the IAEA in force. Signed but not ratified the African Nuclear Weapon-Free Zone Treaty (Pelindaba Treaty).

Chemical weapons and protective equipment
Alleged continued research and possible production of chemical warfare agents. Alleged stockpile of chemical agents (mustard and nerve agents). Personal protective equipment; Soviet type decontamination units; Fuchs (Fox) ABC detection vehicle (12); SPW-40 P2Ch ABC detection vehicle (small quantity).

Refused to sign the CWC.

Biological weapons
Suspected biological warfare program; no details available.

Not a party to the BWC.

Ballistic Missiles

Model	Launchers	Missiles	Since	Notes
SS-1 (Scud B/Scud C)	24	100	1973	Possibly some upgraded
Future procurement				
Scud C/Project-T		90		Locally produced
Vector				Unconfirmed
No-Dong		24		Alleged

Space Assets

Model	Type	Notes
Satellites		
NILESAT-1/2	Communication	Civilian
Ground stations		
Aswan	Remote sensing	Receiving and processing satellite images for desert research
Future procurement		
DesertSat	Environmental	Monitoring coastal erosion, desertification, and water resources
Remote sensing	Reconnaissance	100kg; a sun-synchronous, 668km orbit

Armed Forces

Order-of-Battle

Year	1999	2000	2001	2002	2003
General data					
Personnel (regular)	450,000	450,000	450,000	450,000	450,000
SSM launchers	24	24	24	24	24
Ground forces					
Divisions	12	12	12	12	12
Total number of brigades	49	49	49	49	49
Tanks	~2,750	~3,000	~3,000	~3,000	~3,000
	(3,505)	(3,505)	(3,585)	(3,585)	(3,605)
APCs/AFVs	~3,400	~3,400	~3,400	~3,400	~3,400
	(~5,300)	(~5,300)	(~5,300)	(~5,300)	(~5,300)
Artillery	~3,550	~3,530	~3,530	~3,530	~3,530
(including MRLs)		(~3,570)	(~3,570)	(~3,570)	(~3,570)
Air force					
Combat aircraft	481 (494)	481 (494)	481 (494)	505 (518)	505 (518)
Transport aircraft	44	44	44	48	48
Helicopters	~225	~225	~225	~225	~225
Air defense forces					
Heavy SAM batteries	109	109	109	109	109
Medium SAM batteries	44	44	44	44	44
Light SAM launchers	50	105	105	105	105
Navy					
Submarines	4	4	4	4	4
Combat vessels	65	64	64	62	62
Patrol craft	104	104	104	104	109

Personnel

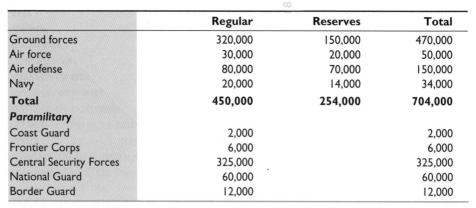

	Regular	Reserves	Total
Ground forces	320,000	150,000	470,000
Air force	30,000	20,000	50,000
Air defense	80,000	70,000	150,000
Navy	20,000	14,000	34,000
Total	**450,000**	**254,000**	**704,000**
Paramilitary			
Coast Guard	2,000		2,000
Frontier Corps	6,000		6,000
Central Security Forces	325,000		325,000
National Guard	60,000		60,000
Border Guard	12,000		12,000

4 IRAN

Major Changes

- Iran admitted it is building a facility for enriching uranium and another one for the production of heavy water. Iran announced that it is developing an indigenous fuel cycle for civilian use but it is generally perceived to be a part of a nuclear weapons program. Iran agreed to accede to the IAEA Additional Protocol and open its facilities to international inspection. It also agreed to suspend its uranium enrichment programs temporarily.
- Although Iran renewed its negotiation with Russia in early 2002 for large weapons deals, none have materialized yet, except for the Mi-17-1 utility helicopters.
- The Shehab-3 MRBM is assessed to be in its early stages of operational status. It is assessed that Iran has some 20 missiles. Iran performed a successful test of the Fateh-110 ballistic missile in late 2002. This is a medium range (200 km) high accuracy solid fuel ballistic missile but its operational status is yet unknown.
- The crash of an Iranian IL-76 aircraft in February 2003 revealed that Iran has been using at least some of the 15 IL-76 transferred from Iraq in 1991. Furthermore, it is possible that Iran activated the ex-Iraqi Su-25 attack aircraft.
- The Iranian armed forces received most of the 55 Mi-17-1 utility helicopters it ordered.
- The Iranian artillery received an unspecified number of Krasnopol 155mm laser guided projectiles.
- The Iranian navy received China Cat fast patrol boats from China as well as C-701 ship-borne missiles from China. The Iranian navy also received 15 small patrol boats from North Korea.

General Data

Official Name of the State: Islamic Republic of Iran
Supreme Religious and Political National Leader (Rahbar): Ayatollah Ali Hoseini Khamenei
Head of State (formally subordinate to National Leader): President Hojatolislam Seyyed Mohammed Khatami
Minister of Defense: Rear Admiral Ali Shamkhani
Commander in Chief of the Armed Forces: Major General Mohammad Salimi
Head of the Armed Forces General Command Headquarters: Major General Hasan Firuzabadi
Chief of the Joint Staff of the Armed Forces: Brigadier General Abdol Ali Pourshasb
Commander of the Ground Forces: Brigadier General Nasser Mohammadi-Far
Commander of the Air Force: Brigadier General Reza Pardis
Commander of the Navy: Rear Admiral Abbas Mohtaj
Commander in Chief of the Islamic Revolutionary Guards Corps (IRGC): Major General Yahya Rahim Safavi
Chief of the Joint Staff of the IRGC: Rear Admiral Ali Akbar Ahmadian
Commander of the IRGC Ground Forces: Brigadier General Aziz Ja'afri
Commander of the IRGC Air Wing: Brigadier General Ahmad Kazemi
Commander of the IRGC Naval Wing: Rear Admiral Ali Morteza Saffari

Area: 1,647,240 sq. km. (not including Abu Musa Island and two Tunb islands; control disputed.)
Population: 69,100,000

Economic Data (in US $billion)

	1998	1999	2000	2001	2002
GDP (current prices)	61.0	55.2	71.9	85.5	114.8
Defense expenditure	1.93	1.64	2.71	4.07	NA

Major Arms Suppliers

Major arms suppliers are Russia, which supplied submarines, MBTs, helicopters, combat and transport aircraft, and AD systems. The PRC supplied fast missile patrol boats, transport aircraft, cruise missiles, and AD systems. North Korea supplied and assisted Iran in the production of SSMs.

Recently the role of Pakistan as supplier of nuclear technology surfaced. Other suppliers include Ukraine, which supplied tanks and transport aircraft, Romania, which supplied AD systems, and France, which supplied trainer aircraft.

Major Arms Transfers

Iran supplied armament and financial aid to Hizbollah in Lebanon. Arms supplies included MRLs, long range rockets, ATGMs, and shoulder-launched SAMs. Some Palestinian organizations received aid that included ATGMs and mortars. Iran cooperated with Syria in the development of ballistic missiles and allegedly in the production of chemical weapons.

Foreign Military Cooperation

Type	Details
Forces deployed abroad	300 IRGC troops in Lebanon
Joint maneuvers	India (1998), Italy (2001), Kuwait (proposed naval maneuvers), Oman (observers 1999), Pakistan (naval maneuvers 2003)
Security agreements	India (2003), Pakistan (2003)

Defense Production

SSMs, tanks, armored combat vehicles, self-propelled guns, towed guns, artillery rockets, anti-tanks missiles, attack helicopters, transport aircraft, trainer aircraft, helicopters, patrol crafts, midget submarines, UAVs, AD systems, cruise missiles, guided bombs, radars, fire control systems.

Note: Some of the weapon systems may be copies of foreign types and not indigenously developed. In addition, some may be only prototypes, which were displayed for propaganda purposes and are not in production.

Strategic Assets

NBC Capabilities

Nuclear capability
One 5 Mw research reactor acquired from the US in the 1960s (in Tehran) and one small 27 kw miniature neutron source reactor (in Esfahan). One 1,000 Mw VVER power reactor under construction, under a contract with Russia, in Bushehr; uranium enrichment facility in Natanz and heavy water production facility in Arak – connected to an alleged nuclear weapons program.

Party to the NPT. Safeguards agreement with the IAEA in force. Recently agreed to sign the IAEA Additional Protocol.

Chemical weapons and protective equipment
Iran admitted in 1999 that it had possessed chemical weapons in the past. Party to the CWC, but nevertheless suspected of still producing and stockpiling mustard, sarin, soman, tabun, VX, and other chemical agents. Alleged delivery systems include aerial bombs, artillery shells, and SSM warheads. PRC and Russian firms and individuals allegedly provide assistance in CW technology and precursors.

Personal protective equipment and munitions decontamination units for part of the armed forces.

Biological weapons
Suspected biological warfare program; no details available.

Party to the BWC.

Ballistic Missiles

Model	Launchers	Missiles	Notes
SS-1 (Scud B/Scud C)	~20	300 Scud B, 100 Scud C	
Shehab-2	+	+	Probably similar to the Syrian Scud-D
Shehab-3	6	~20	
CSS-8	16		
Total	**~40**		
Future procurement			
Fateh-110			Under development

Space Assets

Name	Type	Notes
Ground station		
IRSC	Remote sensing	Multi-spectral remote sensing
Future procurement		
Mesbah	Research	To be launched in 2005 in cooperation with Italy
Zohreh	Communication	Was to be launched in 2003 by Russian SLV. Postponed or cancelled

Armed Forces

Order-of-Battle

Year	1999	2000	2001	2002	2003
General data					
Personnel (regular)	~520,000	~520,000	~520,000	~520,000	~520,000
SSM launchers	~30	~30	~40	~40	~40
Ground forces					
Divisions	32	32	32	32	32
Total number of brigades	87	87	87	87	87
Tanks	1,520	~1,500	~1,700	~1,700	~1,700
APCs/AFVs	1,235	1,240	~1,570*	~1,570	~1,570
Artillery	2,640	~2,700	~2,700	~2,700	~2,700
(including MRLs)	(2,930)	(~3,000)	(~3,000)	(~3,000)	(~3,000)
Air force					
Combat aircraft	205 (297)	205 (333)	209 (337)	207 (335)	207 (342)
Transport aircraft	91 (112)	92 (111)	105 (123)	105 (124)	105 (134)
Helicopters	293 (555)	300 (560)	325 (560)	345 (580)	365 (600)
Air defense forces					
Heavy SAM batteries	30–35	30–35	29*	29	29
Medium SAM batteries	+	+	+	+	+
Light SAM launchers	95	95	95	95	95
Navy					
Submarines	3	3	3	3	3
Combat vessels	31	29	29	29	29
Patrol craft	139	~120	~110	~110	~110

* Due to change in estimate.

Personnel

	Regular	Reserves	Total
Ground forces	~350,000	350,000	700,000
Air force	18,000		18,000
Air defense	12,000		12,000
Navy	~18,000		18,000
IRGC – ground forces	100,000		100,000
IRGC – navy	20,000		20,000
Total	**~518,000**	**350,000**	**868,000**
Paramilitary			
Baseej		2,000,000	2,000,000

5   IRAQ

Major Changes

- The Iraqi armed forces were completely eliminated in the war. All units have been disbanded and large quantities of matériel have been destroyed. No information about the present state of this military materiel is available.
- A large occupation force is still present in Iraq. This force is predominantly American and British but personnel from some 30 countries are present in Iraq. Besides military personnel the CPA employs private companies in both logistical and purely military duties.
- The American administration is trying to organize and train various security organizations. Aside from an army and a civilian police force, these include a civil defense force, a border police, and a facilities protection service. So far the training of the army has been the slowest. Australia will train a coastal guard, and South Africa will train the special forces.
- Although the Iraqi arsenal of weapons of mass destruction was one of the purported justifications for the war against Iraq, to date no such weapons have been uncovered.

General Data

Official Name of the State: The Republic of Iraq
Head of State: President Ghazi al-Yawar
Prime Minister: Iyad Alawi
Minister of Defense: Hazem Shaalan

Area: 431,162 sq. km.
Population: 23,600,000 est.

Economic Data (in US $billion)

	1997	1998	1999	2000	2001
GDP (current prices)	15.3	18.1	23.7	31.8	27.9

Note: Economic data on Iraq is scarce and unreliable.

Major Arms Suppliers

Since 1991 Iraq did not receive any major arms supplies. Small-scale suppliers included Belarus, Russia, Yugoslavia, the Czech Republic, and Ukraine, all of which supplied air defense and radar related equipment.

Defense Production

Since the 2003 war the Iraqi defense industry has not functioned, and it will likely not regain any considerable production capability in the foreseeable future. After the 1991 Gulf War and until the 2003 war, Iraq produced ballistic missiles, UAVs – or converted combat aircraft to weapon-carrying UAVs, and upgraded its air defense systems.

Strategic Assets

NBC Capabilities

Nuclear capability

Since the 1991 war, all known Iraqi facilities have been destroyed by the UN and IAEA facility destruction and monitoring teams. Renewed inspections after the 2003 war did not reveal any renewal of the nuclear weapons program.

Party to the NPT.

Chemical weapons and protective equipment

Inspections by American teams following the 2003 war did not reveal any evidence of renewed production of chemical agents or weapons.

Chemical agents produced in the past included mustard (sulfur mustard and purified mustard), sarin, tabun, soman, VX, hydrogen cyanide (unconfirmed); large quantities of chemical agents were destroyed by UN missions, but some may have remained.

Delivery systems: SSM warheads, L-29 converted UAVs, artillery shells, mortar bombs, MRL rockets, aerial bombs, and land mines. To date, no delivery systems are known to exist.

Personal protective equipment; Soviet-type unit decontamination equipment.

Not a party to the CWC.

Biological weapons

Inspections by American teams following the 2003 war did not reveal any evidence of renewed production of biological agents or weapons.

Biological agents produced in the past included anthrax, aphlatoxin, botulinum, and typhoid. Iraq claims that they were destroyed, but stocks were largely unaffected by UN inspection activity. Experiments were also carried out with other agents.

Delivery systems: SSM warheads, aerial bombs, and airborne spraying-tanks for combat aircraft, helicopters, and UAVs. To date, no delivery systems are known to exist.

Party to the BWC.

Ballistic Missiles

Following the 1991 Gulf War, Iraq produced the al-Soumoud and the al-Ababil short-range ballistic missiles. In addition Iraq was suspected of hiding some al-Hussayn missiles. Since the 2003 war the ballistic missile capability has been eliminated and probably will not be renewed in the foreseeable future.

Armed Forces

Order-of-Battle

Year	1999	2000	2001	2002	2003
General data					
Personnel (regular)	432,500	432,500	432,500	432,500	5,560
SSM launchers*	5	5	5	10	0
Ground forces					
Divisions	23	23	23	23	0
Tanks	2,000	2,000	2,000	2,000	0
	(2,300)	(2,400)	(2,400)	(2,400)	
APCs/AFVs	2,000	2,000	2,000	2,000	0
	(2,900)	(2,900)	(2,900)	(2,900)	
Artillery	2,050	2,100	2,100	2,100	0
(including MRLs)					
Air force					
Combat aircraft	215 (333)	215 (333)	200 (333)	120 (333)	0
Transport aircraft	+	+	+	+	0
Helicopters	370 (460)	370 (460)	370 (460)	370 (460)	0
Air defense forces					
Heavy SAM batteries	60	60	60	60	0
Medium SAM batteries	NA	NA	10	10	0
Light SAM launchers	130	130	130	130	0
Navy					
Combat vessels	2	0	0	1	0
Patrol craft	0	0	0	1	0

* Number does not include unguided rocket launchers.

Personnel

	Active	Anticipated
Iraqi National Army	5,560	35,000
Special Forces		6,500
Civil Defense Corps	35,000	40,000
Facility Protection Service	74,000	80,000
Coastal Defense Force		
Border Police	8,780	8,835
Border Enforcement Guards	10,000	16,000

Foreign Military Personnel

Approximately 130,000 US troops. International coalition forces includes Australia (1,000), Azerbaijan (150), Bulgaria (480), Czech Republic (80), Denmark (496), Dominican Republic (300), El Salvador (360), Japan (250), Jordan (400), Hungary (300), Italy (2,700), Latvia (120), Lithuania (100), Mongolia (180), Norway (150), Netherlands (1,100), New Zealand (60), Philippines (100), Poland (2,400), Portugal (128), Romania (700), Singapore (200), Slovakia (100), South Korea (700), Thailand (460), Turkey (4,000), UK (11,000), Ukraine (2,000).

6 ISRAEL

Major Changes

- While the Israeli army is still absorbing its Merkava Mk III MBTs, the new Merkava Mk IV has been presented and has begun entering into service.
- The Israeli army merged two divisional HQs that dealt with security in the West Bank into a single HQ. The change is due to budgetary constraints.
- The Israel air force received its first Sufa (F-16I) combat aircraft. The air force chose to implement its option on the F-16I, and ordered 52 additional aircraft. These will be supplied between 2005 and 2008.
- The first Gulfstream V aircraft ordered by the air force arrived in Israel, where they will be equipped with electronic equipment for SIGINT missions. The rest of the ordered aircraft are scheduled to arrive in Israel this year.
- The air force deployed the second Arrow BMD battery.
- Meanwhile the air force is considering phasing out older aircraft and closing down some of its squadrons – due to budget cuts. In addition, the Hawkeye E-2C early warning planes have been deactivated.
- The air force began using the new GROB-120B (Snunit) trainer aircraft. They are operated by a private company for the air force flying school.
- Israel launched its new reconnaissance satellite Ofeq 5 to replace the aging Ofeq 3. Besides the military Ofeq series, Israel operates the EROS 1A, which is the first of a projected system of eight civilian reconnaissance satellites. The Eros 1B is scheduled to be launched in 2004.
- The Israeli navy ordered eight new patrol boats – six Super Devora Mk II and two Shaldag.

General Data

Official Name of the State: State of Israel
Head of State: President Moshe Katsav
Prime Minister: Ariel Sharon
Minister of Defense: Shaul Mofaz
Chief of General Staff: Lieutenant General Moshe Ya'alon
Commander of the Air Force: Major General Eliezer Shkedi
Commander of the Army HQ: Major General Yiftah Ron Tal
Commander of the Navy: Rear Admiral Yedidia Ya'ari

Area: 22,145 sq. km., including East Jerusalem and its vicinity, and the Golan Heights.
Population: 6,700,000

Economic Data (in US $billion)

	1999	2000	2001	2002	2003
GDP (current prices)	103.8	114.8	112.7	103.7	109.0
Defense expenditure	8.43	8.93	9.03	9.84	8.72

Major Arms Suppliers

The US is Israel's major foreign arms supplier, supplying combat aircraft, training aircraft, attack heli-copters, helicopters, missile corvettes, tank transporters, SAMs, JDAMs, naval SSMs, MLRS, ATGMs, AMRAAM, SP artillery, and other systems.

Other suppliers include Germany, which supplied Dolphin submarines, training aircraft, NBC detection vehicles, CW protection gear, and Seahake heavy torpedoes. South Africa supplied patrol boats, the Netherlands supplied CW protection gear and assistance in building patrol boats, France supplied training aircraft and CW detectors, and Canada supplied helicopter simulators.

Major Arms Transfers

India is the major Israeli arms receiver. Recently it ordered three Phalcon AEW aircraft in a nearly $1 billion deal. It procured from Israel UAVs, radars, patrol boats, naval SAMs, anti-radar drones, communication equipment, and surveillance systems.

The US procured AGMs, AAMs, digital mapping systems, airborne search and rescue systems, tactical air-launched decoys, flight simulators, mortars, central computers for AFVs, and mine clearing systems. Turkey received AGMs, debriefing systems, aircraft simulators, radars, ECMs, anti-radar missiles, search and rescue systems, and debriefing systems.

Other recipients include Angola (aircraft for ELINT and surveillance, transport helicopters), Argentina (radars and reconnaissance systems), Australia (ESMs, APCs radars, night vision equipment, guns for patrol boats), Belgium (UAVs, debriefing systems), Brazil (combat aircraft, avionics suits), Canada (ESMs, OWSs), Chile (patrol boats, AAMs, AAGs, missiles for patrol boats), Cyprus (torpedo boats, radars, flack jackets, ECSs), Ecuador (combat aircraft), Eritrea (patrol boats), Finland (UAVs, ATGMs, communication equipment), France (SAR systems, debriefing systems), Greece (EW systems, patrol boats), Italy (laser guided bombs, SAR system, Litening pods, debriefing systems, simulators), Netherlands (ATGMs, C^2 systems, debriefing systems), Nicaragua (patrol boats), Philippines (mini-UAVs), Poland (ATGMs), Portugal (ESMs, debriefing systems), Romania (OWS-25 systems, ground radar systems), Singapore (debriefing systems, UAVs, ATGMs, reconnaissance satellite), South Korea (EW systems, AGMs, anti-radar drones, night vision systems, debriefing systems, radars, satellite reconnaissance equipment), Spain (radars), Sri Lanka (attack aircraft, MFPBs, UAVs, radars, ESMs, patrol boats), Sweden (radars), Taiwan (submarines), Thailand (mini-UAVs, search and rescue systems), UK (ESMs, debriefing systems), Venezuela (radars, Litening pods, SAMs, ESMs).

Foreign Military Cooperation

Type	Details
Foreign forces	Pre-positioning of $200 million worth of stockpiled US military equipment; Patriot batteries from the US and Germany were deployed during the war with Iraq (2003)
Cooperation in military training	US and Turkish use of Israeli airfields and airspace for training (2003); Israeli use of Turkish airspace and airfields for training (2003)
Joint maneuvers	Jordan – SAR (1998), Turkey – SAR and joint air force maneuvers (2003), US (2003)

Defense Production

Major systems produced by Israel include: Merkava MBTs, THEL system, SP AAGs, ATRLs, simulators, Arrow ATBM, ALCMs, AAMs, AGMs, CBUs, TV and guided bombs, radars, UAVs and mini-UAVs, attack UAV, LCTs, MFPBs, PBs, SSMs, ELINT equipment, ESM, EW jammers, command and control systems, night vision devices, satellite launchers, imaging satellites, communications satellites.

Strategic Assets

NBC Capabilities

Nuclear capability
Two nuclear research reactors; alleged stockpile of nuclear weapons.*

Not a party to the NPT.

Chemical weapons and protective equipment
Personal protective equipment; Unit decontamination equipment.

Fuchs (Fox) NBC detection vehicles (eight vehicles); SPW-40 P2Ch NBC detection vehicles (50 vehicles); AP-2C CW detectors.

Signed but not yet ratified the CWC.

Biological weapons
Not a party to the BWC.

* According to foreign publications, as cited by Israeli publications.

Ballistic Missiles

Model	Launchers	Missiles	Since	Notes
MGM-52C (Lance)	12		1976	
Jericho Mk 1/2/3 SSM*	+			
Total	+			

* According to foreign publications, as cited by Israeli publications.

Space Assets

Model	Type	Notes
Satellites		
Amos 1/2	Communication	Civilian
Ofeq series	Reconnaissance	Currently deployed Ofeq 5
Eros	Reconnaissance	Civilian derivative of Ofeq
TechSat	Research	Civilian
Future launches		
MILCOM	Communication	
TECHSAR	Reconnaissance	
David	Remote sensing	

Armed Forces

Order-of-Battle

Year	1999	2000	2001	2002	2003
General data					
Personnel (regular)	186,500	186,500	186,500	186,500	186,500
SSM launchers	+	+	+	+	+
Ground forces					
Divisions	16	16	16	16	16
Total number of brigades	77	76	76	76	76
Tanks	3,895	3,930	3,930	3,930	3,700 (3,930)
APCs/AFVs	8,040	8,040	8,040	8,000	7,710
Artillery	1,348	1,348	1,348	+	+
(including MRLs)	(1,948)	(1,948)	(1,948)		
Air force					
Combat aircraft	624 (801)	628 (800)	533 (798)	538 (798)	518 (798)
Transport aircraft	77 (87)	77 (87)	64 (87)	64 (79)	58 (63)
Helicopters	289 (299)	287 (297)	232 (297)	239 (302)	205 (283)
Air defense forces					
Heavy SAM batteries	22	22	22	22	22
Light SAM launchers	~70	~70	~70	~70	~70
Navy					
Submarines	4	6	6	5	5
Combat vessels	21	20	20	15	15
Patrol craft	35	32	32	33	33

Personnel

	Regular	Reserves	Total
Ground forces	141,000	380,000	521,000
Air force	36,000	55,000	91,000
Navy	9,500	10,000	19,500
Total	**186,500**	**445,000**	**631,500**
Paramilitary			
Border Police	7,650		7,650

7   JORDAN

Major Changes

- The American assistance to Jordan was increased to $198 million in 2004. In the wake of the war against Iraq, Jordan will also receive special assistance of $1 billion of which $406 million will be military assistance.
- The Jordanian army is in the process of receiving its second order of 114 Challenger I MBT from Great Britain. Challenger I is replacing the Tariq (Centurion) MBTs. Jordan launched a project to upgrade at least 100 of its M60 MBTs with new engines and new fire control systems. The first battalion of these upgraded MBTs has already been equipped.
- Jordan received 100 Ratel 20 IFVs from South Africa and launched a project to upgrade its M113 APCs. The Jordanian military is also introducing into service a variety of locally developed and produced light vehicles.
- The Jordanian air force received eight more F-16 combat aircraft out of an order of 16 aircraft from a USAF drawdown.
- The air academy of the Royal Jordanian Air Force acquired 16 Firefly training aircraft, which will replace the air academy's aging Bulldogs. The air academy is also acquiring two types of locally assembled light aircraft.
- The RJAF is absorbing its new EC-635 light helicopters, which are still being delivered.
- The US army deployed three Patriot SAM batteries in Jordan. The batteries were leased to Jordan and it is possible that they will remain in Jordan in the future.

General Data

Official Name of the State: The Hashemite Kingdom of Jordan
Head of State: King Abdullah bin Hussein al-Hashimi
Prime Minister: Faisal al-Fayez
Minister of Defense: Faisal al-Fayez
Inspector General of the Armed Forces: Major General Abd Khalaf al-Najada
Chief of the Joint Staff of the Armed Forces: Lieutenant General Khalid Sarayrah
Commander of the Air Force: Major General Prince Faisal bin Hussein
Commander of the Navy: Commodore Ali Mahmoud al-Khasawna

Area: 90,700 sq. km.
Population: 5,300,000

Economic Data (in US $billion)

	1998	1999	2000	2001	2002
GDP (current prices)	7.9	8.1	8.5	9.0	9.5
Defense expenditure	0.71	0.75	0.79	0.79	0.78

Major Arms Suppliers

Major arms suppliers include the US, which supplied combat aircraft, self-propelled artillery, anti-tank missiles, and radars, and the UK, which supplied MBTs and training aircraft and upgraded light tanks.

Other suppliers include Ukraine (APCs), Belgium (APCs), France (helicopters), Turkey (transport aircraft), and Canada (upgrading of transport aircraft).

Major Arms Transfers

Jordan sold used combat aircraft to the Philippines and light armored vehicles to Qatar and UAE.

Foreign Military Cooperation

Type	Details
Cooperation in military training	Turkey (use of facilities and airspace for training of pilots); training for the new Iraqi forces
Foreign forces	Some US prepositioning of military equipment
Forces deployed abroad	Small contingency force in Afghanistan, Congo (MONUC), East Timor (UNTAET), Ethiopia and Eritrea (UNMEE); observers in Georgia (UNOMIG) and Tajikstan, Iraq, Kosovo (UNMIK)
Joint maneuvers	Egypt, Israel (SAR 1998), France, Oman, Qatar, Turkey, UAE, UK, US
Security agreements	Saudi Arabia, Turkey, US

Defense Production

Upgrading of tanks, APCs, conversion of IFVs, night vision equipment.

Strategic Assets

NBC Capabilities

Nuclear capability
No known capability.

Party to the NPT.

Chemical weapons and protective equipment
No known CW activities.

Personal protective and decontamination equipment.

Party to the CWC.

Biological weapons
No known BW capability.

Party to the BWC.

Armed Forces

Order-of-Battle

Year	1999	2000	2001	2002	2003
General data					
Personnel (regular)	94,200	94,200	94,200	100,700	100,700
Ground forces					
Divisions	4	4	4	4	4
Total number of brigades	14	14	14	14	14
Tanks	~900	~900	~920	~990	~970
	(~1,200)	(~1,200)	(~1,270)	(~1,442)	(~1,467)
APCs/AFVs	1,475	1,500	1,500	1,606	1,813
	(1,575)	(1,600)	(1,750)	(1,806)	(2,013)
Artillery	788	788	838	838	844
(including MRLs)	(813)	(813)	(863)	(863)	(867)
Air force					
Combat aircraft	101	100	91 (100)	91 (100)	97 (106)
Transport aircraft	12 (14)	14 (16)	12	14	14
Helicopters	68	68	74	74	83
Air defense forces					
Heavy SAM batteries	14	14	14	14	17
Medium SAM batteries	50	50	50	50	50
Light SAM launchers	50	50	50	50	50
Navy					
Patrol craft	10	13	13	10	10

Personnel

	Regular	Reserves	Total
Ground forces	88,000	60,000	148,000
Air force	12,000		12,000
Navy	700		700
Total	**100,700**	**60,000**	**160,700**
Paramilitary			
General Security Forces	25,000		25,000
(including Desert Patrol)			
Popular Army		200,000–250,000	200,000–250,000

Note: The Popular Army is not regarded as a fighting force.

8 KUWAIT

Major Changes

- Since the end of the Iraq War, the American forces in Kuwait are in a process of reduction.
- The American administration approved a Kuwaiti request to acquire Apache AH-64D helicopters – some of them will be equipped with the Longbow radar.
- The administration also approved the sale of AMRAAM AIM-120B advanced AA missiles for the Kuwaiti F-18 combat aircraft.
- The Kuwaiti coast guard ordered three patrol boats from Australia.

General Data

Official Name of the State: State of Kuwait
Head of State: Jabir al-Ahmad al-Jabir al-Sabah
Prime Minister: Saad Abdallah al-Salim al-Sabah
Minister of Defense: Jabir Mubarak al-Hamad al-Sabah
Chief of General Staff: Major General Fahd Ahmad al-Amir
Commander of the Air Force and Air Defense Forces: Brigadier General Sabir al-Suwaidan
Commander of the Navy: Commodore Ahmad Yousuf al-Mualla

Area: 17,820 sq. km. (including 2,590 sq. km. of the Neutral Zone)
Population: 2,400,000

Economic Data (in US $billion)

	1998	1999	2000	2001	2002
GDP (current prices)	25.1	29.2	37	34.2	35.3
Defense expenditure	2.27	2.43	3.08	3.65	3.81

Major Arms Suppliers

Kuwait diversifies its arms procurements. Its major suppliers are the US, the UK, and France. The US supplies attack helicopters, ATGMs, fire control radar, AMRAAM, LASS air defense system, communication systems, maintenance aid, contracting and upgrading air bases. France supplies helicopters, MFPBs, SAMs, and anti-ship missiles. The UK supplies C⁴I systems, APCs, SAMs, and anti-ship missiles.

Other suppliers include Germany (NBC reconnaissance vehicles) and Egypt (air defense missiles).

Major Arms Transfers

Brazil (used combat aircraft).

Foreign Military Cooperation

Type	Details
Foreign forces	Coalition forces: since the end of the war there has been a process of force reduction, as some of the forces moved into Iraq and others left the region altogether. As of March 2004, 26,000 US troops remained in Kuwait.
Joint maneuvers	Czech Republic (2003), Egypt (2001), France (2004), GCC countries (2001), Germany (2002), Iran (1998), Jordan (2001), UK (marines), US (amphibious, command post, and naval exercises) (2003)
Security agreements	Belarus (2001), France, GCC countries, Iran (2002), Italy (2003), PRC, Russia, South Africa (2002), UK, US

Strategic Assets

NBC Capabilities

Nuclear capability
No known nuclear activity.

Party to the NPT.

Chemical weapons and protective equipment
Fuchs (Fox) ABC detection vehicles (11). Personal protective equipment; unit decontamination equipment. No known CW activities.

Party to the CWC.

Biological weapons
No known BW activities.

Party to the BWC.

Armed Forces

Order-of-Battle

Year	1999	2000	2001	2002	2003
General data					
Personnel (regular)	19,500	19,500	19,500	15,500	15,500
Ground forces					
Number of brigades	6	6	6	7	7
Tanks	318 (483)	318 (483)	318 (483)	318 (483)	318 (483)
APCs/AFVs	436 (715)	~490 (755)	~530 (797)*	~530 (797)	~530 (797)
Artillery	75 (128)	~70 (~125)	~100	~100	~100
(including MRLs)			(~150)	(~130)	(~155)
Air force					
Combat aircraft	40 (59)	40 (59)	40 (59)	40 (59)	39 (58)
Transport aircraft	5	5	5	5	5
Helicopters	24–27	~25	23 (28)	25 (30)	25 (30)
Air defense forces					
Heavy SAM batteries	12	12	12	12	12
Navy					
Combat vessels	6	10	10	10	10
Patrol craft	51	69	69	69	77

* Due to change in estimate.

Personnel

	Regular	Reserves	Total
Ground forces	11,000	24,000	35,000
Air force	2,500		2,500
Navy	2,000		2,000
Total	**15,500**	**24,000**	**39,500**
Paramilitary			
National Guard	5,000		5,000
Civil Defense	2,000		2,000

9　LEBANON

Major Changes

- No major change was recorded in the Lebanese order-of-battle.
- The Syrian military presence in Lebanon was reduced to 15,000 troops. The UNIFIL force deployed in South Lebanon was reduced as well, to some 2000 troops.
- Lebanon is also the home base for the Hizbollah militia, which continues to operate a large number of MRLs, including Iranian-made, long range Fajr-3, and Fajr-5 rockets, and various shoulder-launched SAMs.

General Data

Official Name of the State: Republic of Lebanon
Head of State: President Emile Lahoud
Prime Minister: Rafiq al-Hariri
Minister of Defense: Khalil Hrawi
Commander in Chief of the Armed Forces: Lieutenant General Michel Sulayman
Chief of General Staff: Brigadier General Fady Abu-Shakra
Commander of the Air Force: Brigadier General George Shaàban
Commander of the Navy: Rear Admiral George Maàlouf

Area: 10,452 sq. km.
Population: 3,700,000

Economic Data (in US $billion)

	1999	2000	2001	2002	2003
GDP (current prices)	16.8	16.4	16.2	17.0	18.2
Defense expenditure	0.57	0.88	0.92	0.81	NA

Major Arms Suppliers

Lebanon has not received any major arms systems since 1998, when it received helicopters and APCs from the US.

Other suppliers are Norway and the Czech Republic, which supplied mine clearing equipment. Iran and Syria supplied artillery rockets and other equipment to the Hizbollah militias.

Foreign Military Cooperation

Type	Details
Foreign forces	Syria (15,000 in Beka', Tripoli area, and Beirut); Palestinian organizations; 300 Iranian Islamic Revolution Guards Corps (IRGC); several instructors with Hizbollah non-government militia in the Syrian-held Beka', UNIFIL forces in South Lebanon (2,000 from France, Ghana, Italy, India, Poland, and Ukraine); Pakistan (300 troops in de-mining operations); UAE (30 army engineers in de-mining operations)

Strategic Assets

NBC Capabilities

Nuclear capability
No nuclear capability.

Party to the NPT.

Chemical weapons and protective equipment
No known CW activities.

Not a party to the CWC.

Biological weapons
No known BW activities.

Party to the BWC.

Armed Forces

Order-of-Battle

Year	1999	2000	2001	2002	2003
General data					
Personnel (regular)	51,400	51,400	51,400	61,400	61,400
Ground forces					
Number of brigades	13	13	13	13	13
Tanks	280 (350)	280 (350)	280 (350)	280 (350)	280 (350)
APCs/AFVs	730	1,235*	1,235	1,235	1,235
	(875)	(1,380)	(1,380)	(1,380)	(1,380)
Artillery (including MRLs)	~330	~330	~335	~335	~335
Air force					
Combat aircraft	(16)	(16)	(6)	(6)	(6)
Transport aircraft	(2)	(1)	(1)	(1)	(1)
Helicopters	16 (34)	16 (38)	16 (38)	16 (38)	16 (38)
Navy					
Patrol craft	39 (41)	32 (35)	32 (35)	20*	20

* Due to change in estimate.

Personnel

	Regular	Reserves	Total
Ground forces	60,000		60,000
Air force	1,000		1,000
Navy	400		400
Total	**61,400**		**61,400**
Paramilitary			
Gendarmerie/internal security	13,000		13,000

10 LIBYA

Major Changes

- Libya announced its willingness to dismantle all its WMD programs. It announced its willingness to sign the IAEA Additional Protocol, and acceded to the CWC. It also allowed US and UK inspectors to enter its WMD facilities. During these inspections it was revealed that Libya had a preliminary clandestine uranium enrichment program. All elements of Libya's WMD have been removed. This included equipment of its nuclear and chemical weapons programs as well as Scud C ballistic missiles.
- Libya probably sold some of its MiG-21 combat aircraft to Uganda.
- Libya also sold its entire fleet of G-222 transport aircraft to a private company, and its entire fleet of CH-47 helicopters to the UAE.

General Data

Official Name of the State: The Great Socialist People's Libyan Arab Jamahiriya
Head of State: Colonel Muammar al-Qaddafi
Prime Minister: Mubarak Abdallah al-Shamikh (official title: Secretary-General of the General People's Committee)
Minister of Defense: Colonel Abu-Bakr Yunis Jaber
Inspector General of the Armed Forces: Colonel Mustapha al-Kharrubi
Commander in Chief of the Armed Forces: Colonel Abu-Bakr Yunis Jaber
Chief of Staff: Brigadier General Ahmed Abdallah Awn
Commander of the Air Force and Air Defense Forces: Brigadier General Ali Riffi al-Sharif

Area: 1,759,540 sq. km.
Population: 5,600,000

Economic Data (in US $billion)

	1999	2000	2001	2002	2003
GDP (current prices)	30.5	34.0	27.5	16.8	17.7
Defense expenditure	NA	NA	NA	NA	NA

Major Arms Suppliers

Libya did not procure major arms systems in the past decade.

Arms suppliers include Ukraine, which supplied transport aircraft and SSMs, and North Korea, which supplied SSMs. Assistance for Libya's weapons program was received mainly from Pakistan, but also from the PRC, Iran, Russia, Ukraine, and private companies in Western Europe.

Foreign Military Cooperation

Type	Details
Forces deployed abroad	Since the end of 2001, about 200 Libyan soldiers have been stationed in the Central African Republic (2002)
Security agreements	Algeria (2001), Italy (2003), Tunisia (2001)

Defense Production

Libya produced toxic chemical agents and SSMs. These capabilities were given up in 2003.

Strategic Assets

NBC Capabilities

Nuclear capability
5Mw Soviet-made research reactor at Tadjoura; Libya had a clandestine uranium enrichment program with a few thousand centrifuges. These were surrendered and removed in the framework of its steps to renounce its WMD programs.

Party to the NPT. Safeguards agreement with the IAEA in force. Signed but not ratified the African Nuclear Weapon-Free Zone Treaty (Treaty of Pelindaba).

Chemical weapons and protective equipment
CW production facilities, stockpile of chemical agents, nerve gas, and mustard gas. In the framework of its steps to renounce its WMD programs, work has been carried out to dismantle all past chemical weapons stockpiles. Libya also acceded to the CWC.

Personal protective equipment; Soviet type decontamination units.

Biological weapons
Alleged production of toxins and other biological weapons (unconfirmed).

Party to the BWC.

Ballistic Missiles

Model	Launchers	Missiles	Since	Notes
Scud B	80	500	1976/1999	
Scud C				Scud C missiles have been removed
Total	~80			

Armed Forces

Order-of-Battle

Year	1999	2000	2001	2002	2003
General data					
Personnel (regular)	76,000	76,000	76,000	76,000	76,000
SSM launchers	128	80*	~80	~80	~80
Ground forces					
Number of brigades	1	1	1	1	1
Number of battalions	46	46	46	46	46
Tanks	600–700	~650	~650	~650	~650
	(2,210)	(2,210)	(2,210)	(2,210)	(2,210)
APCs/AFVs	~2,750	~2,750	~2,750	~2,750	~2,750
	(2,970)	(2,970)	(2,970)	(2,970)	(2,970)
Artillery	2,220	~2,270	~2,320	~2,320	~2,320
(including MRLs)	(2,300)	(~2,350)	(~2,400)	(~2,400)	(~2,400)
Air force					
Combat aircraft	~360 (443)	~360 (443)	~360 (443)	~340 (443)	~328 (431)
Transport aircraft	85 (90)	85 (90)	85 (90)	85 (90)	68 (73)
Helicopters	127 (204)	127 (204)	127 (204)	127 (204)	112 (189)
Air defense forces					
Heavy SAM batteries	~30	~30	~30	~30	~30
Medium SAM batteries	~10	~10	~10	~10	~10
Light SAM launchers	55	55	55	55	55
Navy					
Submarines	0 (4)	0 (4)	0 (4)	0 (2)	0 (2)
Combat vessels	34	24	24	20	20
Patrol craft	2	0	0	0	0

* Due to change in estimate.

Personnel

	Regular	Reserves	Total
Ground forces	50,000		50,000
Air force and air defense	18,000		18,000
Navy	8,000		8,000
Total	**76,000**		**76,000**
Paramilitary			
People's Militia	40,000		40,000
Revolutionary Guards	3,000		3,000
(part of the People's Militia)			
Islamic Pan African Legion	2,500		2,500
(part of the People's Militia)			

11 MOROCCO

Major Changes

- The Moroccan navy received two Floreal frigates from France, along with two Panther naval helicopters.
- No other major change was recorded in the Moroccan armed forces.

General Data

Official Name of the State: Kingdom of Morocco
Head of State: King Mohammed VI
Prime Minister: Driss Jettou
Minister of Defense: King Mohammed VI
Secretary General of National Defense Administration: Abdel Rahaman Sbai
Commander in Chief of the Armed Forces: King Mohammed VI
Inspector General of the Armed Forces: General Abd al-Kader Loubarisi
Commander of the Air Force: Ali Abd al-Aziz al-Omrani
Commander of the Navy: Captain Muhammad al-Tariqi

Area: 622,012 sq. km., including the former Spanish Sahara.
Population: 29,600,000

Economic Data (in US $billion)

	1999	2000	2001	2002	2003
GDP (current prices)	35.2	34.9	36.1	39.1	48.3
Defense expenditure	1.42	1.38	1.39	NA	NA

Major Arms Suppliers

Most of the Moroccan arsenal comes from France and the US. France supplied frigates, patrol boats, and helicopters, and the US supplied training aircraft and APCs.

Other suppliers include Belarus (tanks), and the UK (upgrading of artillery guns).

Foreign Military Cooperation

Type	Details
Forces deployed abroad	Small contingency force in Bosnia and Croatia
Joint maneuvers	France, UK (2003), US (2003)
Security agreements	Tunisia

Strategic Assets

NBC Capabilities

Nuclear capability
No nuclear capability.

Request for nuclear research reactor approved by US government.

Party to the NPT.

Chemical weapons and protective equipment
No known CW activity.

Party to the CWC.

Biological weapons
No known BW activities.

Signed but not ratified the BWC.

Armed Forces

Order-of-Battle

Year	1999	2000	2001	2002	2003
General data					
Personnel (regular)	145,500	145,500	145,500	145,500	145,500
Ground forces					
Number of brigades	6	6	6	6	6
Tanks	379	540*	640	640	640
APCs/AFVs	1,074	1,120*	1,120	1,120	1,120
	(1,374)	(1,420)	(1,420)	(1,420)	(1,420)
Artillery	967	1,027*	1,060	1,060	1,060
(including MRLs)	(1,017)				
Air force					
Combat aircraft	72	72	59 (72)	59 (72)	59 (72)
Transport aircraft	43	43	41 (43)	41 (43)	41 (43)
Helicopters	130	130	121 (131)	121 (131)	122 (132)
Air defense forces					
Light SAM launchers	37	37	37	37	37
Navy					
Combat vessels	13	13	13	13	15
Patrol craft	48	52	52	52	52

* Due to change in estimate.

Personnel

	Regular	Reserves	Total
Ground forces	125,000	150,000	275,000
Air force	13,500		13,500
Navy and marines	7,000		7,000
Total	**145,500**	**150,000**	**295,500**
Paramilitary			
Gendarmerie Royale	10,000		10,000
Force Auxiliere	25,000		25,000
Mobile Intervention Corps	5,000		5,000

12   OMAN

Major Changes

- The Omani air force ordered twelve F-16 C/D combat aircraft. They will be delivered beginning 2005. With these advanced aircraft Oman will acquire advanced armament, including PANTERA target acquisition pods, AMRAAM AAMs, Harpoon anti-ship missiles, and JDAM precision guided munitions.
- The air force ordered 16 Super Lynx helicopters from Britain. These helicopters are scheduled to arrive beginning this year.
- Oman received its 20 Cougar Enforcer patrol boats. These boats will serve the Omani police.
- The Omani navy will also receive twelve small fast patrol boats from Abu Dhabi.

General Data

Official Name of the State: Sultanate of Oman
Head of State: Sultan Qabus ibn Said al-Said
Prime Minister: Sultan Qabus ibn Said al-Said
Minister of Defense: Badr bin Saud bin Harib al-Busaidi
Chief of General Staff: Lieutenant General Ahmad bin Harith bin Naser al-Nabhani
Commander of the Air Force: Air Vice Marshal Yahya bin Rashid al-Juma'ah
Commander of the Navy: Rear Admiral Salim bin Abdalla bin Rashid al-Alawi

Area: 212,000 sq. km.
Population: 2,600,000

Economic Data (in US $billion)

	1999	2000	2001	2002	2003
GDP (current prices)	15.71	19.73	19.94	20.07	21.02
Defense expenditure	1.78	2.10	2.42	2.49	2.44

Major Arms Suppliers

Major arms suppliers include the US, which supplied combat aircraft, sea launch missiles, AAMs, early warning network, and aerial reconnaissance systems; and the UK, which supplied MBTs, missile corvettes, APCs, and air defense radars.

Other suppliers include Switzerland (training aircraft), Spain (patrol boats), Pakistan (training aircraft), Italy (helicopters), Netherlands (surveillance radar), and France (air defense systems).

Foreign Military Cooperation

Type	Details
Foreign forces	Coalition forces: since the end of the war there has been a process of force reduction, as some of the forces moved into Iraq and others left the region altogether. This process is still currently underway.
Joint maneuvers	Egypt (2001), GCC countries (2003), India (2003), Jordan (2001), Pakistan (2002), UK (2001), US (2001)
Security agreements	GCC countries, India (2003), Iran (2003), Turkey (2001), US, Yemen (2004)

Strategic Assets

NBC Capabilities

Nuclear capability
No known nuclear activity.

Signatory to the NPT.

Chemical weapons and protective equipment
No know CW activities.

Party to the CWC.

Biological weapons
No known BW activities.

Party to the BWC.

Armed Forces

Order-of-Battle

Year	1999	2000	2001	2002	2003
General data					
Personnel (regular)	34,000	34,000	34,000	34,000	34,000
Ground forces					
Number of brigades	4	4	4	4	4
Total number of battalions	18	18	18	18	18
Tanks	131 (181)	131 (181)	151 (201)	151 (201)	151 (201)
APCs/AFVs	135 (166)	~135 (~165)	~225* (~335)	~225 (~335)	~385 (~415)
Artillery	148 (154)	148 (154)	148 (154)	148 (154)	148 (154)
Air force					
Combat aircraft	31 (47)	31	29 (30)	29 (30)	29 (30)
Transport aircraft	38 (42)	38 (42)	41 (45)	41 (45)	41 (45)
Helicopters	37	35	41	41	46
Air defense forces					
Light SAM launchers	58	58	58	58	58
Navy					
Combat vessels	9	9	9	9	9
Patrol craft	23	22	22	17	37

* Due to change in estimate.

Personnel

	Regular	Reserves	Total
Ground forces	25,000		25,000
Air force	5,000		5,000
Navy	4,000		4,000
Total	**34,000**		**34,000**
Paramilitary			
Tribal force (Firqat)	3,500		3,500
Police/border police (operating aircraft, and PBs)		7,000	
Royal Household (including Royal Guard, Royal Yachts, and Royal Flight)			6,500

13   PALESTINIAN AUTHORITY

Major Changes

- Ahmad Qurei (Abu Ala) was appointed as prime minister to replace Mahmoud Abbas (Abu Mazen).

General Data

This section includes information on the Palestinian Authority and Palestinian security organizations inside Palestinian Authority territory. It does not cover Palestinians living elsewhere.

Official Name: Palestinian National Authority (PA)
Chairman: Yasir Arafat
Prime Minister: Ahmad Qurei (Abu Ala)
Minister of Internal Security: Hakam Bal'awi
Chief of Security Forces: General Abd al-Rizk al-Majaida

Area: 400 sq. km. (Gaza), 5,800 sq. km. (West Bank).
Population: Gaza: 1,120,000 est.; West Bank: 2,000,000 est.

Economic Data (in US $billion)

	1998	1999	2000	2001	2002
GDP (current prices)	4.23	4.28	4.36	4.4	3.5
Defense expenditure	0.30	0.50	NA	NA	NA

Major Arms Suppliers

The Palestinian forces smuggle arms from Egypt and Lebanon. Sources of these weapons are not always known. In 2001 Iran tried to send a shipload of arms, which was intercepted by Israel.

Defense Production

Palestinian forces produce Qassam rockets, mortars, and explosive charges. Palestinian forces announced they managed to produce ATGMs but these are probably unguided rockets.

Security Forces

Order-of-Battle

Year	1999	2000	2001	2002	2003
General data					
Personnel (regular)	~34,000	~36,000	~45,000	~45,000	~45,000
Ground forces					
APCs/AFVs	45	45	~40	+	+
Artillery				+	+
Aerial Police					
Helicopters	2 (4)	2 (4)	2 (4)	0	0
Coastal Police					
Patrol craft	13	10	13	0	0

Personnel

	Gaza	West Bank	Total	Notes
General Security Service branches				
Public security	+	+	14,000	Also referred to as the National Security Force
Coastal police	+	+	1,000	
Aerial police	+	+	+	Rudimentary unit operating VIP helicopters
Civil police	+	+	10,000	Civilian Police – a law enforcement agency; operates the 700-strong rapid deployment special police
Preventive Security Force	+	+	5,000	Plainclothes internal security force
General Intelligence	+	+	3,000	Intelligence gathering organization
Military Intelligence	+	+	+	Unrecognized preventive security force; includes the Military Police
Civil Defense	+	+	+	Emergency and rescue service
Additional security forces				
Presidential Security	+	+	3,000	Elite unit responsible for Arafat's security
Special Security Force	+	+	+	Unrecognized intelligence organization
Total	**~25,000**	**~20,000**	**~45,000**	

Note: More than three years of armed conflict between the PA and Israel have changed the situation described by this table. Most of the fighting was conducted by non-governmental organizations like the Tanzim and Hamas. At present there is no data concerning the status of the organizations listed above. Some of them ceased to exist, and some may reappear as strong organizations, depending on the personal status of their leaders. Thus, we prefer to display statistics as they were at the end of 2000. The Palestinian security services included several organizations under the "Palestinian Directorate of Police Force" recognized in the Cairo and Washington agreements. In addition, there were some organizations that reported directly to Arafat. Some of the security organizations (particularly the civilian police) have little or no military significance. They are mentioned here because of the unusual organizational structure, and because it is difficult to estimate the size of the total forces that do have military significance.

14 QATAR

Major Changes

- Qatar has become a major base for US activities in the Gulf. Recently the command post of the central command was located in Qatar. The US is operating a large new air base in al-Udeid, in Qatar, besides the large logistics base in al-Sahiliya.
- The Qatari navy ordered four DV-15 fast patrol boats from France. They are scheduled to arrive in 2004.

General Data

Official Name of the State: State of Qatar
Head of State: Shaykh Hamad ibn Khalifa al-Thani
Prime Minister: Abdallah Ibn Khalifa al-Thani
Minister of Defense: Shaykh Hamad ibn Khalifa al-Thani
Commmander in Chief of the Armed Forces: Shaykh Hamad ibn Khalifa al-Thani
Chief of General Staff: Brigadier General Hamad bin Ali al-Attiyah
Commander of the Ground Forces: Colonel Saif Ali al-Hajiri
Commander of the Air Force: General Ali Saeed al-Hawal al-Marri
Commander of the Navy: Captain Said al-Suwaydi

Area: 11,437 sq. km.
Population: 600,000

Economic Data (in US $billion)

	1999	2000	2001	2002	2003
GDP (current prices)	12.4	17.8	17.1	17.5	21.0
Defense expenditure	NA	NA	NA	NA	NA

Major Arms Suppliers

France is the major arms supplier to Qatar. It supplied combat aircraft and MBTs.

The US built major installations, including a large air base and storage facilities. These installations are currently used by US forces.

Foreign Military Cooperation

Type	Details
Forces deployed abroad	Troops part of GCC "Peninsula Shield" rapid deployment force in Saudi Arabia
Foreign forces in country	Coalition forces: since the end of the war there has been a process of force reduction, as some of the forces moved into Iraq and others left the region altogether. This process is still underway.
Joint maneuvers	France (2002), GCC countries (2002), Italy (2000), UK, US (2002), Yemen
Security agreements	Bahrain, France (1994), GCC defense pact (2000), Italy (2001), Iran, Kuwait, Oman (1999), Saudi Arabia

Strategic Assets

NBC Capabilities

Nuclear capability
No known nuclear activity.

Party to the NPT.

Chemical weapons and protective equipment
No known CW activities.

Party to the CWC.

Biological weapons
No known BW activities.

Party to the BWC.

Armed Forces

Order-of-Battle

Year	1999	2000	2001	2002	2003
General data					
Personnel (regular)	11,800	11,800	11,800	11,800	11,800
Ground forces					
Number of brigades	1	2	2	2	2
Number of regiments	1				
Total number of battalions	10	11	11	11	11
Tanks	44	44	44	44	44
APCs/AFVs	222 (302)	~260 (338)	~260 (338)	~260 (338)	~260 (338)
Artillery	56	56	56	56	56
(including MRLs)					
Air force					
Combat aircraft	14	18	18	18	18
Transport aircraft	8	8	7 (8)	7 (8)	7 (8)
Helicopters	31	31	30 (31)	30 (31)	30 (31)
Air defense forces					
Light SAM launchers	48	48	51	51	51
Navy					
Combat vessels	7	7	7	7	7
Patrol craft	36	26	26	13*	13

* Due to change in estimate.

Personnel

	Regular	Reserves	Total
Ground Forces	8,500		8,500
Air Force	1,500		1,500
Navy (including marine police)	1,800		1,800
Total	**11,800**		**11,800**
Paramilitary			
Armed police	8,000		8,000

15   SAUDI ARABIA

Major Changes

- The US withdrew most of its military presence from Saudi Arabia. Currently some 950 American military personnel remain stationed in Saudi Arabia.
- The Saudi navy commissioned its three F-3000 La Fayette frigates. The last one (the al-Dammam) was commissioned in January 2004.

General Data

Official Name of the State: The Kingdom of Saudi Arabia
Head of State: King Fahd ibn Abd al-Aziz al-Saud
Prime Minister: King Fahd ibn Abd al-Aziz al-Saud
First Deputy Prime Minister and Heir Apparent: Crown Prince Abdallah ibn Abd al-Aziz al-Saud
Defense and Aviation Minister: Prince Sultan ibn Abd al-Aziz al-Saud
Chief of General Staff: General Salih ibn Ali al-Muhaya
Commander of the Ground Forces: Lieutenant General Husein al-Qubeel
Commander of the National Guard: Crown Prince Abdallah ibn Abd al-Aziz al-Saud
Commander of the Air Force: Lieutenant General Abd al-Rahman ibn Fahd al-Faisal
Commander of the Navy: Vice Admiral Fahd ibn Abdallah

Area: 2,331,000 sq. km.
Population: 24,200,000

Economic Data (in US $billion)

	1999	2000	2001	2002	2003
GDP (current prices)	161.2	188.7	186.5	188.5	211.6
Defense expenditure	17.6	20.0	25.86	21.56	NA

Major Arms Suppliers

The major arms suppliers to Saudi Arabia are the US and France. The US supplied combat aircraft, AAMs, surveillance radars, anti-tank missiles, and early warning networks. France supplied combat vessels.

Other suppliers include Germany (helicopters), Italy (helicopters), UK (hovercraft, ARM missiles), and Canada (APCs, IFVs).

Foreign Military Cooperation

Type	Details
Foreign forces	GCC "Peninsula Shield" rapid deployment force: 7,000–10,000 men at Hafr al-Batin, mostly Saudis, and from other GCC countries; some 8,000 instructors and technicians from Pakistan. The US has withdrawn most of its military presence in Saudi Arabia. Current American presence includes some 950 personnel.
Joint maneuvers	Egypt (2003), France, GCC countries (2001), Jordan (2001), Pakistan (2004), UK, US (2001)

Defense Production

APCs, radar subsystems, parts of EW equipment.

Strategic Assets

NBC Capabilities

Nuclear capability
No known nuclear activity.

Party to the NPT.

Chemical weapons and protective equipment
No known CW activities.

Personal protective equipment; decontamination units; US-made CAM chemical detection systems; Fuchs (Fox) NBC detection vehicles.

Party to the CWC.

Biological weapons
No known BW activities.

Party to the BWC.

Ballistic Missiles

Model	Launchers	Missiles	Since	Notes
CSS-2	8–12	30–50	1988	Number of launchers unconfirmed

Space Assets

Model	Type	Notes
Satellites		
Arabsat	Communication	Civilian
SaudiSat	Remote sensing	Two (10 kg. each) were launched in
1A/1B/1C	and space research	September 2000 by a Russian military rocket, and are orbiting 650 km. above earth. The third satellite was launched in December 2002.
Ground Stations		
SCRS	Imagery	Receiving SPOT, Landsat, and NOAA

Armed Forces

Order-of-Battle

Year	1999	2000	2001	2002	2003
General data					
Personnel (regular)	165,000	171,500	171,500	171,500	171,500
SSM launchers	8–12	8–12	8–12	8–12	8–12
Ground forces					
Number of brigades	20	20	20	20	20
Tanks	865	750	750	750	750
	(1,015)	(1,015)	(1,015)	(1,015)	(1,015)
APCs/AFVs	~5,310	~5,300	~4,500	~4,630	~4,630
	(~5,440)	(~5,440)	(~5,300)	(~5,430)	(~5,430)
Artillery	~410	~410	~410	~410	~410
(including MRLs)	(~780)	(~780)	(~780)	(~780)	(~780)
Air force					
Combat aircraft	~345	~355	~360	~345	~289*
			(~365)		(~344)
Transport aircraft	61	61	42 (55)*	42 (55)	42 (55)
Helicopters	160	160	214 (216)*	214 (216)	214 (216)
Air defense forces					
Heavy SAM batteries	22	22	25	25	25
Medium SAM batteries	16	16	21	21	21
Navy					
Combat vessels	24	24	24	25	27
Patrol craft	80	74	74	64	68

* Due to change in estimate.

Personnel

	Regular	Reserves	Total
Ground forces	75,000		75,000
Air force	20,000		20,000
Air defense	4,000		4,000
Navy (including a marine unit)	13,500		13,500
National Guard	57,000	20,000	77,000
Royal Guard	2,000		2,000
Total	**171,500**	**20,000**	**191,500**
Paramilitary			
Mujahidun (affiliated with National Guard		30,000	30,000
Coast Guard	4,500		4,500
Frontier Corps	10,500		10,500

16   SUDAN

Major Changes

- The Sudanese air force procured twelve Mig-21 combat aircraft from Ukraine, in addition to the several far more advanced Mig-29 aircraft that the Sudanese air force acquired two years ago.
- The Sudanese army received 30 BTR-80 APC from Russia.

General Data

Official Name of the State: The Republic of Sudan
Head of State: President Omar Hassan Ahmad al-Bashir
Minister of Defense: Major General Bakri Hassan Sallah
Chief of General Staff: General Abbas Arabi
Commander of the Air Force: Major General Ali Mahjoub Mardi
Commander of the Navy: Commodore Abbas al-Said Othman

Area: 2,504,530 sq. km.
Population: 32,540,000

Economic Data (in US $billion)

	1998	1999	2000	2001	2002
GDP (current prices)	10.27	10.04	12.93	14.09	13.50
Defense expenditure	0.21	0.24	0.33	NA	NA

Major Arms Suppliers

Major arms sales to Sudan come from Russia, which supplied combat aircraft. Previously Sudan received arms from Iran, which allegedly supplied tanks, aircraft, other vehicles, and EW equipment.

Other suppliers included Belarus (combat aircraft), Lithuania (helicopters), and Poland (tanks).

Foreign Military Cooperation

Type	Details
Foreign forces	PRC (alleged presence of forces for the defense of Chinese-operated oil fields) (2001)
Forces deployed abroad	Central African Republic (2002)
Security agreements	Syria (2000), Egypt (2001), Russia (2002)

Strategic Assets

NBC Capabilities

Nuclear capability

No known nuclear activity.

Party to the NPT.

Chemical weapons and protective equipment

Alleged CW from Iran (unsubstantiated); alleged production of CW (unsubstantiated); personal protective equipment; unit decontamination equipment

Party to the CWC.

Biological weapons

No known BW activities.

Party to the BWC.

Armed Forces

Order-of-Battle

Year	1999	2000	2001	2002	2003
General data					
Personnel (regular)	103,000	103,000	103,000	104,000	104,000
Ground forces					
Divisions	9	9	9	9	9
Total number of brigades	61	61	61	61	61
Tanks	~320	~350	~350	~350	~350
APCs/AFVs	~560	~560	~560	~545	~575
	(~700)	(~700)	(~700)	(~745)	(~745)
Artillery	~760	~760	~760	~770	~778
(including MRLs)	(~770)	(~770)	(~770)	(~785)	(~793)
Air force					
Combat aircraft	~35 (~55)	~35 (~55)	~35 (55)	~35 (55)	~51 (71)
Transport aircraft	26	25	24	24	24
Helicopters	~60 (69)	~60 (69)	57 (73)	59 (71)	75 (87)
Air defense forces					
Heavy SAM batteries	5	5	20*	20	20
Navy					
Patrol craft	22	18	18	18	16

* Due to change in estimate.

Personnel

	Regular	Reserves	Total
Ground forces	100,000		100,000
Air force	3,000		3,000
Navy	1,000		1,000
Total	**104,000**		**104,000**
Paramilitary			
People's Defense Forces	15,000	85,000	100,000
Border Guard	2,500		2,500

17   SYRIA

Major Changes

- As per a change in estimate, a revised order-of-battle appears below.
- President Bashar Assad appointed Mohammad Ghazi Otri as Syria's new prime minister.
- The Syrian army received a brigade of upgraded T-72 MBTs. The tanks were upgraded by an Italian company and were equipped with a modern, Western fire control system.
- The air defense forces received a quantity of Igla (SA-18) shoulder-launched SAMs.
- Syria tested its first Scud D ballistic missile (with a range of 700km). It is estimated that some of these missiles can be considered operational.

General Data

Official Name of the State: The Arab Republic of Syria
Head of State: President Bashar al-Assad
Prime Minister: Mohammed Jazi Otri
Minister of Defense: Major General Hassan Turkmani
Chief of General Staff: Major General Ali Habib
Commander of the Air Force: Major General Kamal Makhafut
Commander of the Navy: Vice Admiral Wa'il Nasser

Area: 185,180 sq. km.
Population: 17,100,000

Economic Data (in US $billion)

	1998	1999	2000	2001	2002
GDP (current prices)	17.1	17.7	19.4	19.8	20.4
Defense expenditure	1.0	1.0	1.07	1.27	1.36

Major Arms Suppliers

Russia was the major supplier, and it remains the major potential supplier. In the past it supplied Syria with all its major armament systems. Recent arms deals included combat aircraft and ATGMs.

Other suppliers include Iran (ballistic missile technology), North Korea (ballistic missiles), PRC (ballistic missiles), Ukraine (upgrading of tanks, radars), Italy (upgrading of tanks), and Armenia (upgrading of tanks).

Major Arms Transfers

Lebanon (artillery rockets)

Foreign Military Cooperation

Type	Details
Forces deployed abroad	15,000 in Beka', northern Lebanon (Tripoli area), and Beirut

Defense Production

Ballistic missiles, artillery rockets, upgrading of tanks.

Strategic Assets

NBC Capabilities

Nuclear capability
Basic research. Alleged deal with Russia for a 24 Mw reactor. Deals with China for a 27 kw reactor and with Argentina for a 3 Mw research reactor are probably cancelled.

Party to the NPT. Safeguards agreement with the IAEA in force.

Chemical weapons and protective equipment
Stockpiles of nerve gas, including sarin, mustard, and VX.

There are unconfirmed allegations that Syria received Iraq's stockpile of chemical weapons just before the Iraq War broke out.

Delivery vehicles include chemical warheads for SSMs and aerial bombs.

Personal protective equipment; Soviet-type unit decontamination equipment.

Not a party to the CWC.

Biological weapons
Biological weapons and toxins (unconfirmed).

Signed but not ratified the BWC.

Ballistic Missiles

Model	Launchers	Missiles	Since	Notes
SS-1 (Scud B)	18	200	1974	
SS-1 (Scud C)	8	80	1992	
SS-21 (Scarab)	18		1983	
Scud D	+		2002	
Total	~45			

Space Assets

Name	Type	Notes
Satellite imaging GORS	Remote sensing	Using images from Cosmos, ERS, Landsat, SPOT satellites

Armed Forces

Order-of-Battle

Year	1999	2000	2001	2002	2003
General data					
Personnel (regular)	380,000	380,000	380,000	380,000	289,000*
SSM launchers	44	44	44	~45	~45
Ground forces					
Divisions	12	12	12	12	12
Total number of brigades	67	67	67	67	67
Tanks	3,700	3,700	3,700	3,700	3,700
	(4,800)	(4,800)	(4,800)	(4,800)	(4,800)
APCs/AFVs	4,980	~5,000	~5,000	5,060	5,060
Artillery	2,575	~2,600	~2,600	~2,600	~2,600
(including MRLs)	(2,975)	(~3,000)	(~3,000)	(~3,000)	(~3,000)
Air force					
Combat aircraft	520	520	490	490	450* (490)
Transport aircraft	23 (25)	23 (25)	23 (25)	23	23
Helicopters	295	295	285	225*	225
Air defense forces					
Heavy SAM batteries	108	108	108	108	108
Medium SAM batteries	65	64	64	64	64
Light SAM launchers	55	55	55	55	55
Navy					
Submarines	0 (3)	0 (3)	0 (3)		
Combat vessels	24 (27)	14*	14	14	16*
Patrol craft	8	8	8	8	8

* Due to change in estimate.

Personnel

	Regular	Reserves	Total
Ground forces	215,000	100,000	315,000
Air force	30,000	10,000	40,000
Air defense	40,000	20,000	60,000
Navy	4,000	2,500	6,500
Total	**289,000**	**132,500**	**421,500**
Paramilitary			
Gendarmerie	8,000		8,000
Workers' Militia		400,000	400,000

18   TUNISIA

Major Changes

- The Tunisian navy received one additional Kondor I patrol craft. The navy is to receive its first Landing vessel from Italy.
- No other major change was recorded for the Tunisian armed forces.

General Data

Official Name of the State: The Republic of Tunisia
Head of State: President Zayn al-Abedine Bin Ali
Prime Minister: Mohamed Ghannouchi
Minister of Defense: Dali Jazi
Secretary of State for National Defense: Chokri Ayachi
Commander of the Ground Forces: Brigadier General Rashid Amar
Commander of the Air Force: Major General Rida Hamuda Atar
Commander of the Navy: Commodore Brahim Barak

Area: 164,206 sq. km.
Population: 9,800,000

Economic Data (in US $billion)

	1999	2000	2001	2002	2003
GDP (current prices)	20.8	19.5	20.0	21.0	24.8
Defense expenditure	0.36	0.32	0.32	NA	NA

Major Arms Suppliers

Tunisia had no major arms deals in the past decade. Minor acquisitions were from France (APCs), US (transport aircraft), and Italy (transport aircraft).

Foreign Military Cooperation

Type	Details
Forces deployed abroad	Bosnia and Herzegovina (UNMIBH), Congo (MONUC)
Joint maneuvers	France, Greece (2003), Spain (unconfirmed), US
Security cooperation	Egypt (2001), Germany (2003), Greece (2001), Libya (2001), Morocco (2000)

Defense Production

Patrol boats

Strategic Assets

NBC Capabilities

Nuclear capability
No known nuclear activity.

Signatory to the NPT.

Chemical weapons and protective equipment
No known CW activities.

Party to the CWC.

Biological weapons
No known BW activities.

Party to the BWC.

Armed Forces

Order-of-Battle

Year	1999	2000	2001	2002	2003
General data					
Personnel (regular)	35,500	35,500	35,500	35,500	35,500
Ground forces					
Number of brigades	5	5	5	5	5
Tanks	139 (144)	139 (144)	139 (144)	139 (144)	139 (144)
APCs/AFVs	316	316	316	326	326
Artillery (including MRLs)	205 (215)	205 (215)	205 (215)	205	205
Air force					
Combat aircraft	12	12	18	18	18
Transport aircraft	10 (11)	10 (11)	9 (11)	15 (17)	15 (17)
Helicopters	40	44	51	49	47
Air defense forces					
Light SAM launchers	73	83	83	83	83
Navy					
Combat vessels	11	9	9	9	9
Patrol craft	36	37	37	35	36

Personnel

	Regular	Reserves	Total
Ground forces	27,000		27,000
Air force	4,000		4,000
Navy	4,500		4,500
Total	**35,500**		**35,500**
Paramilitary			
Gendarmerie	2,000		2,000
National Guard	7,000		7,000

19   TURKEY

Major Changes

- The Turkish land forces signed a contract with Israel to upgrade 170 M60 MBT - out of some 900 fit for this upgrade.
- The land forces are absorbing a second batch of 551 ACVs from the Turkish industry. The deal is to be completed by the end of 2004.
- An ongoing project is the renewal of the Turkish artillery with 300 Firtina 155mm self-propelled guns and 400 Panther 155mm towed howitzers, all indigenously assembled. These weapon systems began entering service.
- The Turkish air force received all of its upgraded Phantom F-4 combat aircraft. Another upgrade project involves the aging F-5s which are being upgraded and will be used for training.
- The navy and the coast guard received six of nine CN-235 maritime patrol aircraft. Their mission specific systems will be installed separately.

General Data

Official Name of the State: Republic of Turkey
Head of State: President Ahmet Necdet Sezer
Prime Minister: Recep Tayyip Erdogan
Minister of National Defense: Vecdi Gonul
Chief of General Staff: General Hilmi Ozkok
Commander of the Ground Forces: General Aytac Yalman
Commander of the Air Force: General Ibrahim Firtina
Commander of the Navy: Admiral Ozden Ornek

Area: 780,580 sq. km.
Population: 71,300,000

Economic Data (in US $billion)

	1999	2000	2001	2002	2003
GDP (current prices)	184.9	199.3	145.6	183.1	228.2
Defense expenditure	9.95	10.0	7.21	9.22	NA

Major Arms Suppliers

Turkey's major arms suppliers are the US, France, and Germany. The US supplied combat vessels, combat aircraft, helicopters, early warning aircraft, AD missiles, anti-tank missiles, and radars. France supplied combat vessels, helicopters, training aircraft, and cruise missiles. Germany supplied submarines and combat vessels.

Other suppliers include Israel (upgrading of combat aircraft, upgrading of tanks, upgrading of helicopters, anti-radiation drones), South Korea (self-propelled artillery guns), Italy (helicopters, radars), Norway (cruise missiles), Spain (transport aircraft), and the UK (AD missiles).

Major Arms Transfers

Turkey sold armament systems to several countries, including Malaysia (IFVs), UAE (IFVs), Israel (APCs), Jordan (transport aircraft), Azerbaijan (patrol boats, APCs), Croatia (transport aircraft), Georgia (patrol boats, helicopters), Kazakhstan (patrol boats, APCs), and Macedonia (combat aircraft).

Foreign Military Cooperation

Type	Details
Cooperation in training	Albania, Azerbaijan, Israel (mutual use of airspace and training facilities), Jordan (mutual use of airspace and training facilities; joint training of infantry), Georgia, PRC
Forces deployed abroad	Afghanistan (270 troops in ISAF), Albania, Bosnia (1,000 troops in UNMIBH), Cyprus (30,000 troops), East Timor (UNTAET), Georgia (UNOMIG), northern Iraq (1,000 troops), Kosovo (UNMIK), Israel (TIPH), Macedonia (140 troops)
Foreign forces	US (75 troops and 10 KC-135 aircraft)
Joint maneuvers	Albania (naval – 2000), Bulgaria (2003), Georgia (naval), Germany (2003), Israel (2003), Jordan, Macedonia (part of multinational peacekeeping brigade – 2000), Pakistan, Poland, Romania (part of multinational peacekeeping brigade – 2000), US (2003)
Security agreements	Bosnia (2003), Croatia (2001), France, Georgia (2001), Kazakhstan (2000), Lativa, Oman (2001), Pakistan (2003), Syria (2002)

Defense Production

Turkey has a large and diversified defense industry. Its aerospace industry produces or assembles combat aircraft, helicopters, and transport and training aircraft. Its naval industry produces or assembles submarines, combat vessels, and patrol boats. Land based systems produced include ACVs, self-propelled guns, and AD systems. Other munitions produced include artillery rockets and anti-tank missiles. The electronic industry produces radars, electronic warfare systems, and fire control systems.

Strategic Assets

NBC Capabilities

Nuclear capability
One 5Mw TR-2 research reactor at Cekmerce and one 250 kw ITV-TRR research reactor at Istanbul. Turkey intends to order a 1,000 Mw reactor. As a member of NATO, nuclear weapons were deployed in Turkey in the past, and might be deployed there again.

Party to the NPT. Safeguards agreement with the IAEA in force.

Chemical weapons and protective equipment
Personal protective suits; portable chemical detectors; Fox detection vehicles.

Party to the CWC.

Biological weapons
No known BW activity.

Party to the BWC.

Ballistic Missiles

Model	Launchers	Missiles	Since	Notes
ATACMS	12	72	1997	Using MLRS launchers
Future procurement				
J project			2001	Under development

Space Assets

Model	Type	Notes
Ground stations		
BILTEN	Remote sensing	Receiving imagery from Bilsat
SAGRES	Remote sensing	Receiving imagery from SPOT, ERS, RADARSAT and NOAA
Satellites		
Turksat-2A	Communication	Both civilian and military
Bilsat	Remote sensing	120 kg. payload, 686 km. orbit, 12 m. resolution earth observation civilian satellite
Satellite imagery		
Ikonos	Reconnaissance	Commercial satellite imagery
Ofeq 5	Reconnaissance	Sharing of Israeli satellite imagery

Armed Forces

Order-of-Battle

Year	1999	2000	2001	2002	2003
General data					
Personnel (regular)	633,000	633,000	610,000	515,500	515,500
SSM launchers	12	12	12	12	12
Ground forces					
Divisions	5	5	5	3	3
Total number of brigades	67	67	67	63	63
Tanks	4,115	4,205	2,600	2,600	2,600
	(4,190)	(4,280)	(4,255)	(4,255)	(4,255)
APCs/AFVs	4,520	5,460	5,460	5,460	5,790
Artillery	4,312	~4,350	~4,350	~4,355	~4,370
(including MRLs)	(4,611)	(~4,650)	(~4,650)	(~4,655)	(~4,670)
Air force					
Combat aircraft	416	485	~445 (465)	~390 (410)	~400 (422)
Transport aircraft	87	92	90 (94)	93 (97)	90 (100)
Helicopters	381	395	407	461	462
Air defense forces					
Heavy SAM batteries	24	24	24	24	24
Light SAM launchers	86	86	86	86	86
Navy					
Submarines	16	15	14	13	12
Combat vessels	51	65	78	83	84
Patrol craft	88	108	103	113	102

Personnel

	Regular	Reserves	Total
Ground forces	402,000	259,000	661,000
Air force	60,000	65,000	125,000
Navy	53,000	55,000	108,000
Total	**515,000**	**379,000**	**894,000**
Paramilitary			
Coast Guard	2,200		2,200
Gendarmerie/National Guard	180,000	50,000	230,000

20   UNITED ARAB EMIRATES (UAE)

Major Changes

- Delivery of Leclerc MBTs and the other vehicles ordered within the framework of this contract was concluded.
- The army received its first Guardian ACVs from Ukraine. The turrets for these ACVs were ordered separately and will be installed indigenously. Similar turrets will be installed on some of the older Scorpion light tanks.
- The Emiri air force concluded the deal to acquire 80 F-16 block 60 combat aircraft (which will be delivered 2004–2008).
- Meanwhile the Emiri air force received its first Mirage 2000-9. The project is for a total of 63 aircraft, about half of them new and half of them upgraded older Mirage 2000-5 already in service.
- The Emiri navy received its first 12m Transportbat 2000 landing craft (out of twelve ordered under the Ghannata project). The navy also continues the process of upgrading its TNC-45 MFPBs.
- The Emiri coast guard received 24 small, locally produced, fast attack boats, and ordered 30 more of them.

General Data

Official Name of the State: United Arab Emirates
Head of State: Shaykh Zayid ibn Sultan al-Nuhayan, Emir of Abu Dhabi
Prime Minister: Shaykh Maktum ibn Rashid al-Maktum, Emir of Dubai
Minister of Defense: Muhammad ibn Rashid al-Maktum
Chief of General Staff: HRH Lieutenant General Muhammad ibn Zayid al-Nuhayan
Commander of the Air Force and Air Defense Forces: Brigadier General Khalid bin Abdullah al-Buainnain
Commander of the Navy: Brigadier General Suhail Shaheen al-Murar

Area: 82,900 sq. km. (estimate)
Population: 4,000,000 (estimate)
Note: The UAE consists of seven principalities: Abu Dhabi, Dubai, Ras al-Khaima, Sharjah, Umm al-Qaiwain, al-Fujairah, and Ajman.

Economic Data (in US $billion)

	1999	2000	2001	2002	2003
GDP (current prices)	57.9	70.2	67.6	71.0	78.4
Defense expenditure	1.64	1.64	1.64	NA	NA

Major Arms Suppliers

The US and France are UAE's major arms suppliers. The US supplied combat aircraft, attack helicopters, AAMs, naval SAMs, anti-ship missiles, command and control aircraft, and advanced air launched munitions. France supplied combat aircraft, helicopters, MBTs, SAMs, ARVs, torpedos, anti-ship missiles, and C³I systems.

Other suppliers include Netherlands (frigates, surveillance radars), Germany (training aircraft, APCs, tank transporters, ABC detection vehicles), Indonesia (maritime patrol aircraft), Romania (upgrading of helicopters), Russia (procurement and upgrading of IFVs, air defense systems), Spain (patrol aircraft), UK (AGMs, sonar), and South Africa (EW systems).

Foreign Military Cooperation

Type	Details
Foreign forces	Some 70 US soldiers (2003)
Forces deployed abroad	In Saudi Arabia (part of GCC "Peninsula Shield" rapid deployment force)
Joint maneuvers	Egypt (2001), France (2002), GCC countries (2002), Jordan (2001), US (2001), Turkey (2002)
Security agreements	France (2000), Germany, India (2003), Slovak Republic (1999)

Defense Production

UAE's industry produces patrol boats, corvettes, amphibious landing craft, and assembles UAVs and mini-UAVs and target drones.

Strategic Assets

NBC Capabilities

Nuclear capability
No known nuclear activity.

Signatory to the NPT.

Chemical weapons and protective equipment
No known CW activities. Personal protective equipment; unit decontamination equipment.

Party to the CWC.

Biological weapons
No known BW activities.

Signed but not ratified the BWC.

Ballistic Missiles

Model	Launchers	Missiles	Since	Notes
Scud B	6		1991	Owned by Dubai; unconfirmed

Space Assets

Model	Type	Notes
Satellites Thuraya-1/2	Communications	Geosynchronous, civilian satellites. The first was launched in September 2000, the second in June 2003
Ground stations Dubai Space Imaging	Remote sensing	Receiving satellite images from Ikonos and India's IRS satellites

Armed Forces

Order-of-Battle

Year	1999	2000	2001	2002	2003
General data					
Personnel (regular)	46,500	46,500	46,500	65,500	65,500
SSM launchers	6	6	6	6	6
Ground forces					
Number of brigades	8	8	8	8	8
Tanks	~330 (~430)	~430 (~470)	~400 (~470)	532 (604)	539 (611)
APCs/AFVs	~960 (~1,120)	~1,250 (~1,400)	~1,250 (~1,410)	~1,190 (~1,350)	~1,190 (~1,350)
Artillery (including MRLs)	411 (434)	399 (422)	399 (422)	399 (422)	405 (428)
Air force					
Combat aircraft	54 (66)	54 (66)	54 (66)	54 (66)	47 (70)
Transport aircraft	31 (34)	31 (34)	33 (36)	33 (36)	33 (36)
Helicopters	93 (95)	100 (102)	91 (103)	102 (114)	102 (114)
Air defense forces					
Heavy SAM batteries	~7	5	5	5	5
Medium SAM batteries	6	6	6	6	6
Light SAM launchers	113	~115	~115	~115	~115
Navy					
Combat vessels	12	12	12	12	12
Patrol craft	105	110	112	112	128

Military Forces

Personnel

	Regular	Reserves	Total
Ground forces	59,000		59,000
Air force	4,500		4,500
Navy	2,000		2,000
Total	**65,500**		**65,500**
Paramilitary			
Coast Guard	+		+
Frontier Corps	+		+

21 YEMEN

Major Changes

- Yemen continues working to improve its defense relations with the US. The US is training Yemeni personnel in counterterrorism activities and helping Yemen to establish its coast guard.
- Yemen has acquired 100 small fast patrol boats from Abu Dhabi.
- The Yemeni navy acquired four landing craft from Poland and ordered ten Bay patrol boats from Australia.

General Data

Official Name of the State: Republic of Yemen
Head of State: President Ali Abdallah Salih
Prime Minister: Abd al-Qadir Ba Jamal
Minister of Defense: Brigadier General Abdallah Ali Alaywa
Chief of General Staff: Brigadier General Abdallah Ali Alaywah
Commander of the Air Force: Colonel Muhammad Salih al-Ahmar
Commander of the Navy: Admiral Abdallah al-Mujawar

Area: 527,970 sq. km.
Population: 20,000,000

Economic Data (in US $billion)

	1999	2000	2001	2002	2003
GDP (current prices)	6.7	8.4	8.4	10.8	11.6
Defense expenditure	0.37	0.4	0.46	NA	NA

Major Arms Suppliers

Russia is Yemen's major arms supplier. It supplied combat aircraft and MBTs.

Other suppliers include North Korea (SSMs), Poland (landing craft), Czech Republic (training aircraft, tanks), Australia (patrol boats), and the US (C³ systems, patrol boats).

Foreign Military Cooperation

Type	Details
Foreign forces	Some 23 US soldiers (2003)
Joint maneuvers	US (2004)
Security agreements	Oman (2004), Turkey (2002)

Strategic Assets

NBC Capabilities

Nuclear capability
No known nuclear activity.

Signatory to the NPT.

Chemical weapons and protective equipment
No known CW activities.

Party to the CWC.

Biological weapons
No known BW activities.

Party to the BWC.

Ballistic Missiles

Model	Launchers	Missiles	Since	Notes
SS-1 (Scud B)	6			New missiles received from North Korea, possibly Scud C
SS-21 (Scarab)	4		1988	
Total	10			

Note: Serviceability of missiles and launchers unknown.

Armed Forces

Note: Since the 1994 civil war, all figures are rough estimates.

Order-of-Battle

Year	1999	2000	2001	2002	2003
General data					
Personnel (regular)	~65,000	~65,000	~65,000	~65,000	~65,000
SSM launchers	10	10	10	10	10
Ground forces					
Number of brigades	33	33	33	33	33
Tanks	575	605	~715	~715	~715
	(1,040)	(1,070)	(~1,180)	(~1,180)	(~1,180)
APCs/AFVs	~480	~480	~480	~495	~495
	(~1,170)	(~1,200)	(~1,200)	(~1,210)	(~1,210)
Artillery	~670	~670	~670	~675	~675
(including MRLs)	(~990)	(~1,000)	(~1,000)	(~1,025)	(~1,025)
Air force					
Combat aircraft	~50	~50	~55	~65	~65
	(~150)	(~150)	(~180)	(~190)	(~190)
Transport aircraft	18 (23)	18 (23)	20 (30)	20 (30)	20 (30)
Helicopters	27 (67)	26 (66)	26 (70)	26 (70)	26 (70)
Air defense forces					
Heavy SAM batteries	25	25	25	25	25
Medium SAM batteries	+	+	+	+	+
Light SAM launchers	120	120	120	120	120
Navy					
Combat vessels	11	10	10	10	10
Patrol craft	3	9	9	8	8

Personnel

	Regular	Reserves	Total
Ground forces	~60,000	200,000	~260,000
Air force	3,000		3,000
Navy	2,000		2,000
Total	**~65,000**	**200,000**	**~265,000**
Paramilitary			
Central Security Force	50,000		50,000

Note: The military forces are a combination of personnel from the former Yemen Arab Republic and the People's Democratic Arab Republic; no information regarding reorganization is available.

Tables and Charts

The Middle East Military Balance at a Glance

State	Personnel			Ground Forces			
	Regular	Reserves	Total	Tanks	Fighting vehilces	Artillery	Ballistic missile launchers
Eastern Mediterranean							
Egypt	450,000	254,000	704,000	~3,000	3,680	~3,530	24
Israel	186,500	445,000	631,500	3,700	7,710	1,348	+
Jordan	100,700	60,000	160,700	~970	1,815	844	
Lebanon	61,400		61,400	280	1,235	~335	
Palestinian Authority	~45,000		~45,000				
Syria	289,000	132,500	421,500	3,700	5,060	2,990	~45
Turkey	515,000	379,000	894,000	2,600	5,788	~4,370	12
Persian Gulf							
Bahrain	7,400		7,400	180	277	48	9
Iran	~520,000	350,000	870,000	1,680	~1,570	~2,700	~40
Iraq	5,560		5,560	0	0	0	0
Kuwait	15,500	24,000	39,500	318	~530	~100	
Oman	34,000		34,000	151	~385	148	
Qatar	11,800		11,800	44	~260	56	
Saudi Arabia	171,500	20,000	191,500	750	~4,630	~410	12
UAE	65,500		65,500	532	~1,200	405	6
North Africa and others							
Algeria	127,000	150,000	277,000	900	2,010	900	
Libya	76,000		76,000	~650	~2,750	~2,320	~80
Morocco	145,500	150,000	295,500	640	1,120	1,060	
Sudan	104,000		104,000	~350	~575	~760	
Tunisia	35,500		35,500	139	326	205	
Yemen	65,000	200,000	265,000	~715	~495	~675	10

The Middle East Military Balance at a Glance (continued)

Air Force			Air Defense			Submarines	Navy	
Combat aircraft	Transport aircraft	Helicopters	Heavy batteries	Medium batteries	Light launchers		Combat vessels	Patrol craft
Eastern Mediterranean								
505	44	~225	109	44	105	4	62	109
518	62	205	23		~70	5	15	33
97	14	83	17	17	50			10
		16						20
451	23	225	108	68	55		16	8
397	90	462	24		86	12	84	102
Persian Gulf								
34	3	40	1	2	40		11	21
207	105	365	30		95	3	28	~110
0	0	0	0		0		0	0
39	5	25	12	1			10	74
29	41	46			58		9	37
18	7	30			51		7	13
286	42	214	25	21			27	68
48	33	102	5	6	~115		12	104
North Africa and others								
228	40	174	11	18	78	2	26	16
329	68	112	30	17	55		20	
59	41	122			37		15	52
50	24	75	20					16
18	15	47			83		9	36
67	20	26	25		120		10	8

Weapons of Mass Destruction

| State | Chemical | Biological | Nuclear | SSM Launchers | | |
				Up to 150 km.	150–600 km.	600– 3,000 km.
Eastern Mediterranean						
Egypt	weapons program	R&D	R&D		24	+
Israel	R&D	R&D	alleged weapons	12		+
Syria	weapons program	weapons program	R&D	18	26	1
Turkey	none	none	R&D	12		
Persian Gulf						
Bahrain	none	none	none	9		
Iran	weapons program	weapons program	R&D	16	20	5
Saudi Arabia	none	none	none			12
UAE	none	none	none		6	
North Africa and others						
Algeria	none	none	R&D			
Libya	weapons program	weapons program	R&D		80	+
Sudan	weapons program	none	none			

Space Assets

State	Imagery ground stations	Communication satellites	Research satellites	Reconnaissance satellites	SLVs
Eastern Mediterranean					
Egypt	+	+			
Israel		+	+	+	+
Syria	+				
Turkey		+	+		
Persian Gulf					
Saudi Arabia	+	+			
UAE	+	+			
Iran			+		
North Africa and others					
Algeria			+		

The Eastern Mediterranean Military Forces

Eastern Mediterranean – Personnel

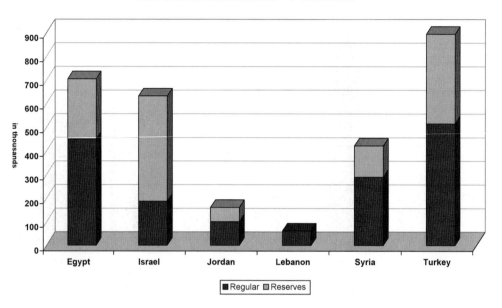

Eastern Mediterranean – Armor

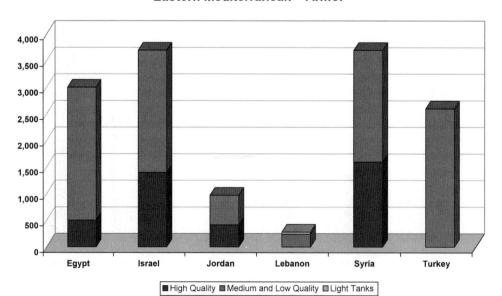

The Eastern Mediterranean Military Forces (continued)

Eastern Mediterranean – ACVs

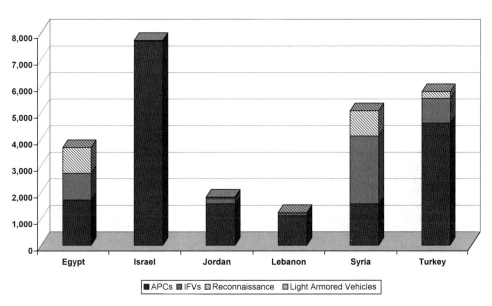

Eastern Mediterranean – Artillery

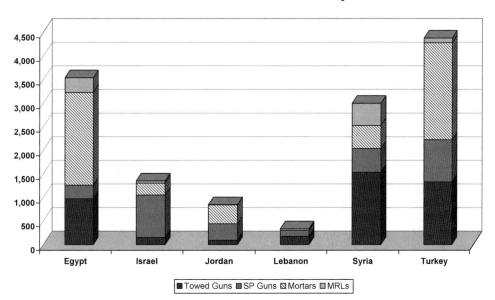

The Eastern Mediterranean Military Forces (continued)

Eastern Mediterranean – Combat Aircraft

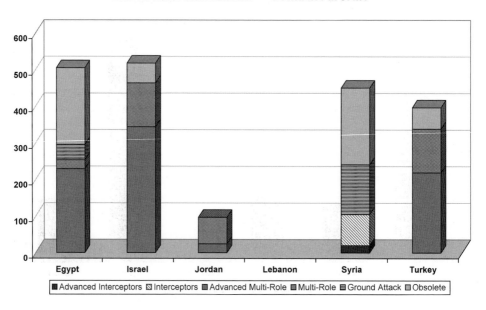

Eastern Mediterranean – Helicopters

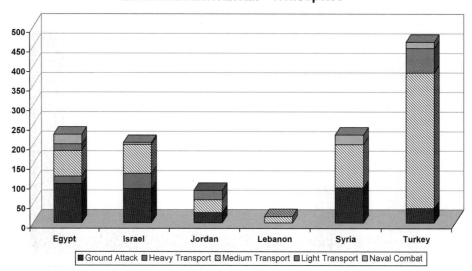

The Eastern Mediterranean Military Forces (continued)

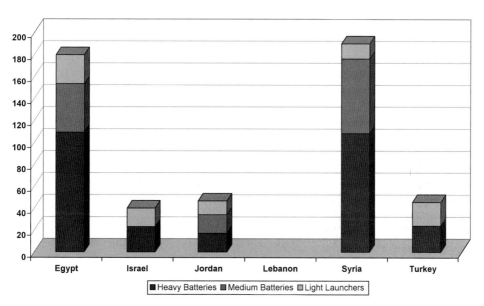

Eastern Mediterranean – Air Defense

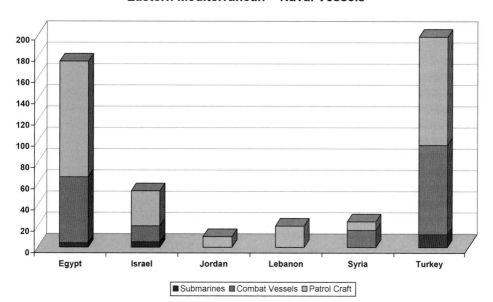

Eastern Mediterranean – Naval Vessels

The Persian Gulf Military Forces

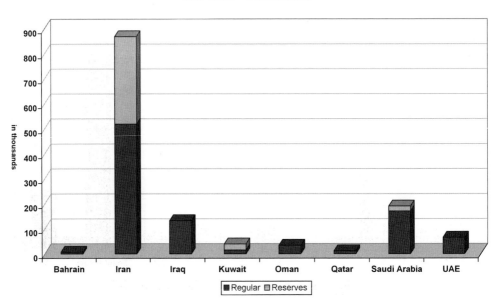

The Gulf – Personnel

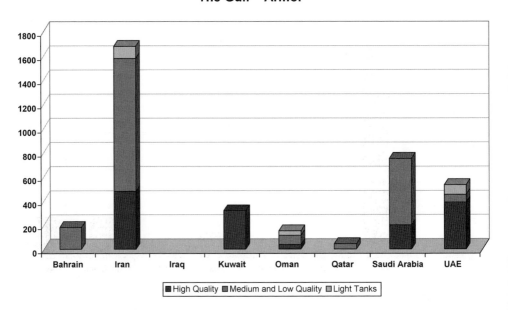

The Gulf – Armor

The Persian Gulf Military Forces (continued)

The Gulf – ACVs

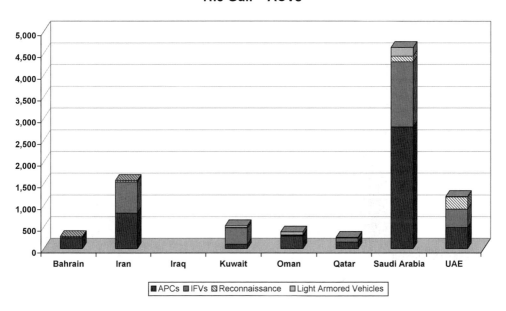

The Gulf – Artillery

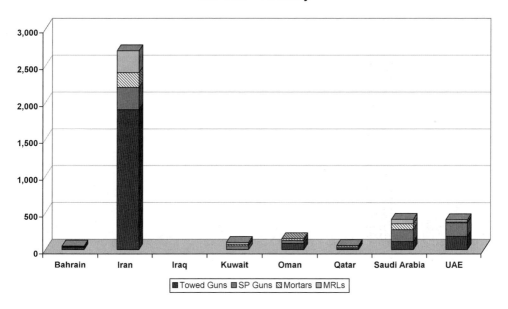

The Persian Gulf Military Forces (continued)

The Gulf – Combat Aircraft

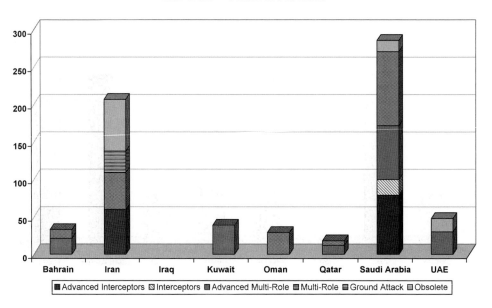

The Gulf – Helicopters

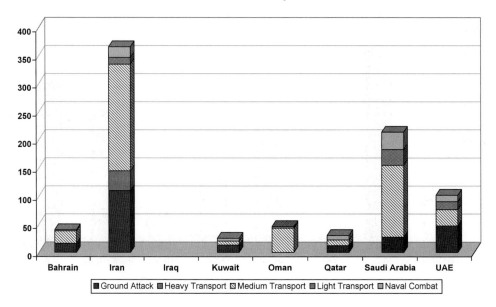

The Persian Gulf Military Forces (continued)

The Gulf – Air Defense

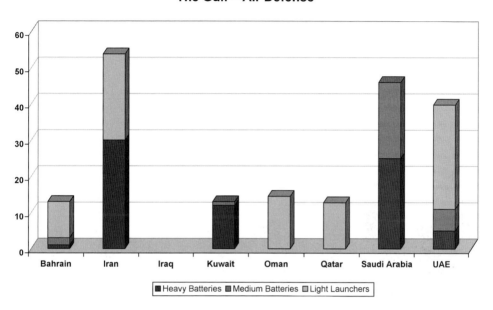

The Gulf – Naval Vessels

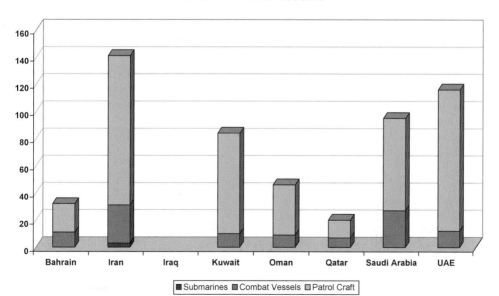

The North African Military Forces

North Africa – Personnel

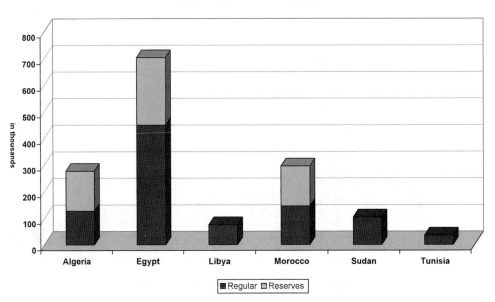

North Africa – Armor

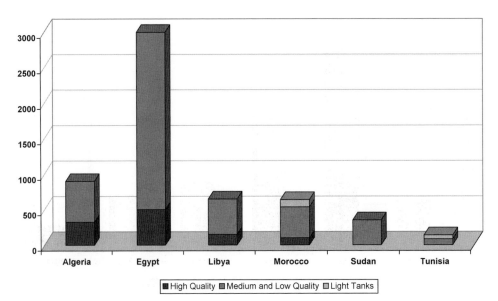

The North African Military Forces (continued)

North Africa – ACVs

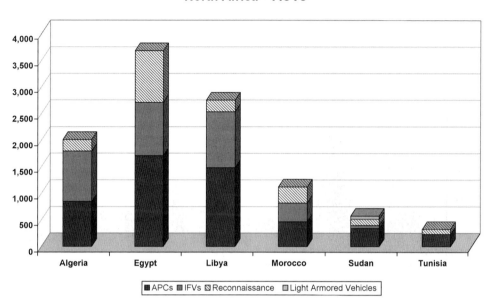

North Africa – Artillery

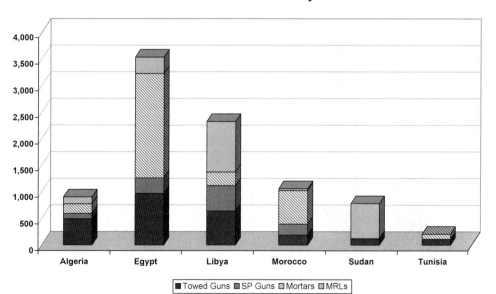

The North African Military Forces (continued)

North Africa – Combat Aircraft

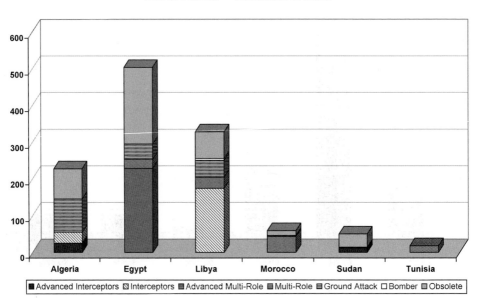

■ Advanced Interceptors ◫ Interceptors ▤ Advanced Multi-Role ▥ Multi-Role ☰ Ground Attack ☐ Bomber ▤ Obsolete

North Africa – Helicopters

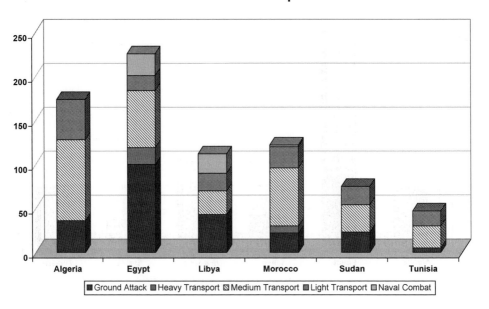

■ Ground Attack ▤ Heavy Transport ▨ Medium Transport ▥ Light Transport ▥ Naval Combat

The North African Military Forces (continued)

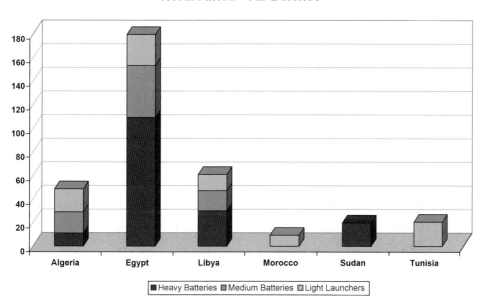

North Africa – Air Defense

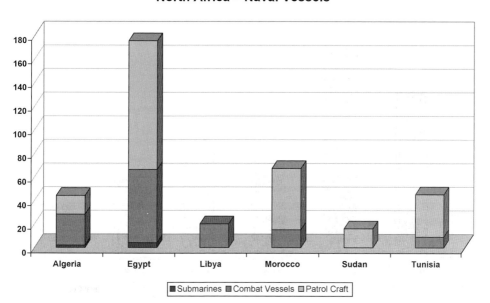

North Africa – Naval Vessels

Contributors

Editors

Shai Feldman was appointed Head of the Jaffee Center in 1997, prior to which he was a Senior Research Associate since the Center's establishment. He is a member of the UN Secretary General's Advisory Board on Disarmament Matters, the Scientific Advisory Committee of the Stockholm International Peace Research Institute (SIPRI), and other organizations. Dr. Feldman has written extensively on nuclear weapons proliferation and arms control in the Middle East, US policy in the region, American-Israeli relations, and the Middle East peace process. Among his recent publications are *After the War in Iraq: Defining the New Strategic Balance* (ed.); *Nuclear Weapons and Arms Control in the Middle East;* and *Bridging the Gap: A Future Security Architecture for the Middle East* (with Abdullah Toukan).

Yiftah S. Shapir joined the Jaffe Center in 1993 as an associate of the Center's Project on Security and Arms Control, where he followed the proliferation of weapons of mass destruction (WMD) in the Middle East. He is head of JCSS's Middle East Military Balance Project, and he is responsible for the quantitative section of the annual *Middle East Strategic Balance.* Shapir served as an officer in the Israeli Air Force, and has extensive background in information technology and operations research.

Other Contributors

Shlomo Brom joined the Jaffee Center as a Senior Research Associate in 1998 after a long career in the IDF. His most senior post in the IDF was Head of the Strategic Planning Division in the Planning Branch of the General Staff. Brig. Gen. Brom participated actively in peace negotiations with the Palestinians, Jordan, and Syria. In 2000 he was named Deputy to the National Security Advisor, returning to JCSS at the end of his post. He is the director of the Jaffee Center's Israel Defense Policy Review project.

Meir Elran joined the Jaffee Center in 2003 after a long career in the IDF Military Intelligence directorate. His most senior post in the IDF was Deputy Director of Military Intelligence (1987-1989). Brig. Gen. Elran participated actively in peace negotiations with Egypt and was a senior member of the Israeli delegation to the peace talks with Jordan. Brig. Gen Elran has served as a senior advisor with several Israeli ministries, where he concentrated on social issues relating to national security.

Ram Erez, a Ph.D. candidate in International Relations at the Hebrew University, joined the Jaffee Center in 2001. His research focuses on how policy networks influence security policy in Israel. At the Jaffee Center, Erez is involved with the arms control project and the project on Israel–European security relations.

Mark A. Heller is Principal Research Associate at the Jaffee Center and editor of *Tel Aviv Notes*. He has been affiliated with the Jaffee Center since 1979 and has taught international relations at Tel Aviv University and at leading universities in the US. Dr. Heller has written extensively on Middle Eastern political and strategic issues. He is also currently a member of the Steering Committee of EuroMeSCo, the Euro-Mediterranean consortium of foreign policy research institutes.

Ephraim Kam, Deputy Head of the Jaffee Center, served as a colonel in the Research Division of IDF Military Intelligence until 1993, when he joined the Jaffee Center. Positions he held in the IDF included Assistant Director of the Research Division for Evaluation and Senior Instructor at the IDF's National Defense College. Dr. Kam specializes in security problems of the Middle East, strategic intelligence, and Israel's national security issues.

Anat Kurz is a Senior Research Associate at the Jaffee Center and has headed the Center's Project on Low-Intensity Conflict since 1989. She has lectured and published extensively on insurgency-related issues, sub-state radical organizations, terrorism as a mode of warfare, and dilemmas of combating terrorism, specializing in Islamic organizations and the Palestinian national movement. Dr. Kurz's current research focuses on the institutionalization of popular struggles.

Emily B. Landau is director of the Jaffee Center's Arms Control and Regional Security Project at JCSS. She has published on CSBMs in the Middle East, Arab perceptions of Israel's qualitative edge, Israeli–Egyptian relations, Israel's arms control policy, and the Arms Control and Regional Security working group of the Madrid peace process (ACRS). Dr. Landau's current research focuses on regional dynamics and processes in the Middle East and developments in arms control thinking.

Paul Rivlin has a joint appointment at the Jaffee Center and at the Moshe Dayan Center for Middle East and African Studies at Tel Aviv University. Dr. Rivlin specializes in the political economies of Arab states, and has published widely on defense economics, oil market trends, and economic development in the Middle East.

Yoram Schweizter joined the Jaffe Center research staff as an expert on international terror. He has lectured and published widely on terror-related issues, and consults for government ministries on a private basis. His areas of expertise include the "Afghan Alumni" phenomenon and the threat posed by Bin Laden, suicide terrorism, and state-sponsored terrorism.